MANAGING WATER AS AN
ECONOMIC RESOURCE

DEVELOPMENT POLICY STUDIES
Edited by John Farrington and Tony Killick
for the Overseas Development Institute, London

This series presents the results of ODI research on policy issues confronted by the governments of developing countries and their partners in aid and trade. It will be of interest to policy makers and practitioners in government and international organisations and to students and researchers in both North and South.

MANAGING WATER AS AN ECONOMIC RESOURCE

James Winpenny

London and New York

First published 1994
by Routledge
2 Park Square, Milton Park, Abingdon, Oxon, OX14 4RN

Transferred to Digital Printing 2005

Simultaneously published in the USA and Canada
by Routledge
270 Madison Ave, New York NY 10016

Phototypeset in Garamond by Intype, London

British Library Cataloguing in Publication Data

A catalogue record for this book is available from the British Library

Library of Congress Cataloging in Publication Data

Winpenny, J. T.
Managing Water as an Economic Resource / James Winpenny.
p. cm.
Includes bibliographical references.
1. Water-supply–Management. 2. Water-supply–Economic aspects.
3. Water resources development. I. Title.
HD1691.W54 1994
333.91–dc20 93–10560

ISBN 0–415–10378–9

CONTENTS

ACKNOWLEDGEMENTS

Much of the material on which this book is based was collected during research carried out for the Economic and Social Committee on Overseas Research (ESCOR) programme of the Overseas Development Administration. My first debt is therefore to the ODA for its professional and financial support for this project, and specifically to Dr Nick Highton for his time and encouragement.

My colleagues at the Overseas Development Institute have also been a plentiful source of advice and encouragement. Tony Killick and Mary Tiffen made helpful detailed comments on successive drafts. Linden Vincent and Peter Ferguson have been abundant sources of material and suggestions.

At an early stage of research I was fortunate in securing the collaboration of the staff of the Water and Sanitation Division of the World Bank, who provided me with facilities and access to the wealth of material in their possession. This collaboration is bearing fruit in the publication of a joint World Bank/ODI Technical Paper, *Policies for Water Conservation and Reallocation: Good Practice Cases in Efficiency and Equity.* I am deeply indebted to John Briscoe, Ramesh Bhatia, Rita Cestti and others for their generous co-operation.

I greatly benefited from Mohan Munasinghe letting me see his new book in draft.

The discussion of Hungary's experience in Chapter 3 was made possible by a Specialist Tour undertaken by the author in October 1991 under the aegis of the British Council and the Hungarian Ministry of Education and Culture. Officials in the Hungarian Ministry of Environment, the VITUKI Water Resources Centre, and the National Water Authority were generous with their time and advice.

Other people who have kindly provided advice or written material are: Tony Allan, Andrew Bennett, Jeremy Berkoff, Chris Bolt, Peter Bolton, Anthony Bottrall, John Dixon, David Durham, Ken Frederick, Paul Herrington, Charles Howe, Brian Jackson, John Makin, James Nickum, Ronan Palmer, Dhira Phantumvanit, Terry Pike, Michel Potier, Robert Repetto, John Richards, Hilary Sunman, David Wall and George Whitlam. I am grateful to them all and hope they are not displeased at the outcome.

Margaret Cornell edited the text and made it more readable.

The lines from T. S. Eliot's *Little Gidding* reproduced at the beginning of Chapter 3 appear by kind permission of the publishers Faber and Faber Ltd and Harcourt Brace & Co.

Finally, I dedicate the book to my wife Lyndsay, and to my children Guy, Helen, Diana, Dominique and Max. If, as scientists claim, water makes up a high proportion of our bodies, it is to be hoped that they recognise something of themselves in this book!

1

THE WATER PROBLEM

day after day, day after day the same –
a weary waste of waters.

(R. Southey, *Madoc*)

This book is written in the belief that water is becoming one of the largest, and certainly the most universal, of problems facing mankind as the earth moves into the twenty-first century. The tasks of supplying enough water of the required quality to growing populations and the safe disposal of wastewater are straining many authorities to the limit. Although the problem varies in type and intensity, it is challenging governments of countries at all stages of development, in most parts of the world.

It will be argued in this chapter that in many cases the failure to treat water as a scarce commodity lies at the heart of the problem. Chapter 2 discusses various approaches to problem-solving, and makes the case for the greater use of 'demand management', whereby better use is made of existing water supplies instead of automatically investing in new supply capacity to satisfy imagined future requirements. Chapter 3 discusses the ingredients of a policy mix which would enable water to be recognised for its true value and better use to be made of existing supplies. In Chapter 4 various examples of demand management are evaluated on different criteria, and are shown to stand up well against conventional supply augmentation.

Although the arguments for the new approach are cogent, it is recognised that there will be powerful political and social resistance to reforms in this most sensitive of sectors. The book concludes therefore (Chapter 5) with a discussion of some implications of the argument for policy-makers, and suggestions for how the required measures could be introduced.

This chapter sets the scene by describing some of the symptoms of the water problem, reviews some underlying causes – revolving around a failure to recognise its economic value – and indicates the grave consequences of this.

SYMPTOMS

The need to conserve water and allocate it to socially more valuable uses has not always been evident. In some societies water has long been treated as a scarce and valuable resource. In the majority of countries, however, water has been treated as though it were available in unlimited quantities, and supplied at zero or low cost to consumers who resent the idea of water as an economic resource. Consumers, abetted by their governments, have clear notions of their water 'requirements', and the task of water authorities has, until recently, been seen as supplying those needs, with cost a secondary consideration. Pricing for water services has been meagre and sporadic, and is normally incidental to cost-recovery, narrowly conceived.

This 'entitlement' syndrome, relying on supply-side solutions to requirements taken as given, is unsustainable in many countries, for a number of pressing reasons – hydrological, environmental and financial.

Hydrological limits

To an increasing extent, easily accessible water sources have already been tapped, supplies are approaching their physical limits, and new supplies for growing populations and rising consumption levels are available only at increasing cost. The syndrome of 'water stress' is widespread. It has been argued that societies suffer water stress when annual renewable supplies fall below approximately 2,000 cubic metres per person at a time when demands for water are increasing in the process of development (Falkenmark, 1989). On this reckoning, by the year 2000 six out of seven East African countries and all North African countries will fall below this critical level (Munasinghe, 1992).

Egypt is near the limit of its surface supplies. In China 50 cities face acute shortages and the groundwater under Beijing is falling by 1–2 metres per year. Tens of thousands of Indian villages currently face shortages. In the Middle East, Israel, Jordan and the West Bank are expected to be using all their renewable sources by 1995 (Postel, 1989). The following case is typical:

> The depletion of freshwater resources for domestic, industrial and agricultural purposes is likely to become the single most important environmental issue facing the Mediterranean countries and, in particular, their coastal areas. Even countries not yet facing water crises are likely to experience large increases in the cost of providing water to meet growing demand. Improving the planning, management and conservation of water will be critical for economic development. Failure to protect freshwater resources will render existing water-based patterns of development unsustainable in a number of countries by significantly increasing the cost of water over time.
>
> (World Bank/EIB, 1990: 26)

The cost of developing new water sources is rising rapidly

as all the readily accessible sources are tapped. A sample of World Bank schemes demonstrates that the cost of producing a unit of water from the 'next project' is often two to three times that of the current one (Bhatia and Falkenmark, 1991).

Conflicts between different classes of user – e.g. farmers, urban and industrial consumers – are becoming more common as scarce supplies are shared out amongst a growing number, and hallowed rights and privileges are increasingly questioned. In the process, farmers, normally the largest water users in dry areas, usually lose out to urban claims, causing animosity and, often, radical social changes. At the international level, rivers, lakes and bays are often shared by several countries; the increased scarcity of water will thus aggravate international tensions in some sensitive regions (e.g. the Middle East and South Asia) (Clarke, 1991). The division of the Nile waters, the use of the Euphrates, the sharing of the Indus, Brahmaputra and Zambesi, and the claims on the Jordan are all sensitive areas, and several of these could become flashpoints before long.

Environmental costs

The environmental costs of water supply schemes are becoming less acceptable as they become greater, and as they are increasingly measured in economic terms. These costs arise both in supply (e.g. depleting aquifers, damming rivers, destroying wetlands) and in the disposal of wastewater (run-off, effluent, sewage) (Winpenny, 1991).

One of the best known environmental catastrophes is the dessication of the Aral Sea in Central Asia. The Sea has shrunk to a fraction of its former volume, depth and area, causing a collapse of the fishing industry, dust storms, and widespread salt deposition (Frederick, 1991). At the same time, the uncontrolled use of agro-chemicals is causing water pollution and serious health problems, including alleged mental retardation amongst children

(Kotlyakov, 1991). These are amongst the true costs of irrigated cotton farming as practised in Uzbekistan and Kazakhstan, some of which can be estimated. If these costs had been better understood at the time, a different course of development would surely have been followed.

The increasing use of techniques for measuring environmental costs has helped to bring home the size of the damage. For example, an extensive literature on environmental values in US wetlands compares their values in alternative and mutually exclusive uses (Leitch and Ekstrom, 1989). This has enabled recreational, hydrological and amenity values to be ranged against the Siren voices of developers and farmers. Proposals to use the waters of the Columbia and Snake Rivers for irrigation and municipal use in other states have aroused the wrath of environmentalists and many others. Economic work on the value of rivers for recreation, amenity and fishing has provided coinage to weigh in this balance (Fisher, 1981). Environmental costs are becoming monetised as the victims of water diversion schemes are increasingly taking to the courts. The legal battles between the Los Angeles Metropolitan Water Authority and the residents of Owens River Valley were among the first and best known, but there have been many others (Reisner, 1990).

Financial constraints

Water utilities, and their governmental sponsors, are in no position to bear the increasing capital, operating and maintenance costs of catering for the projected growth of water requirements. Their poor financial position is partly due to failures in pricing and cost-recovery. In an internal review of World Bank supported projects (probably above-average performers) the effective price per unit of water was found, on average, to be only one-third of the full economic cost of supplying that increment.

Other factors leading to poor financial performance are the high proportion of leaks and wastage, weak billing

and collection systems, and erratic payments by large consumers. These factors interact, to produce a vicious downward spiral. Many systems have a high proportion of
'unaccounted-for water', commonly in the range 25–50
per cent of the amount entering distribution (World Bank,
1992a) – a mixture of physical losses through leakage,
illegal connections and theft, and under-recording of consumption. Add to this the evasion of payment and delays
in collection, and the financial problems of utilities
become apparent. In turn, they are depleted of the funds
necessary to maintain, repair and expand the systems. As
consumers suffer reduced levels of service, their willingness to pay existing bills, let alone increased tariffs, starts
to wilt.

In consequence, many water utilities are badly placed
to make up the backlog of services or to expand the
network to meet the growth of population or of urban
centres. More advanced systems face the additional burden
of coping with higher quality standards and the cost of
replacing antiquated pipes and sewers. As a result:

> By the early 1990s it became apparent that the public
> sector . . . could no longer bear the sole burden of
> investment financing for the sector in industrialized
> countries (such as the United Kingdom), let alone in
> developing countries.
>
> (World Bank, 1992)

'Water stress' symptoms are breaking out everywhere:

> a large irrigation project in India does not operate
> because water has been diverted to the rapidly grow-
> ing city of Pune. In China industries are having to
> reduce their production due to water shortages even
> though they are surrounded by paddy fields. In Cali-
> fornia selenium salts leached by irrigation are killing
> wildlife. [World] Bank irrigation projects in Algeria
> are now competing with Bank urban water supply
> projects for the same water, and many proposed irri-

gation projects and most hydro project proposals are
on hold because of environmental concerns.

(Rogers, 1990)

CAUSES

These symptoms are clear signs that supply systems and
consumption habits have, in general, failed to adapt to the
increasing pressure of demand on the water resource and
to the environmental strains that it causes.

There are many sound reasons why consumption habits
have evolved as they have in different societies. In wetter
parts of the world, the availability of water has been taken
for granted, its provision has been a routine operation,
and engineering solutions have tended to predominate. In
drier regions, water has always been a central preoccu-
pation, and many different kinds of social arrangements
have arisen to deal with the allocation and use of this
scarce resource. Even in humid areas, water shortages
periodically arise (as in the UK since the mid-1980s), and
the control of pollution demands continual vigilance. In
drier areas the age-old problem is everywhere becoming
more serious in the face of population growth and economic development.

There have been institutional, policy and market failures
on a universal scale. This section will highlight three
underlying causes of the problem: the fact that water is
underpriced compared to its real cost of provision, the
fact that water is often a public good which makes it
difficult to extract an economic price from users, and the
existence of environmental 'externalities' in the use of
water which are not reflected in its price.

Under-pricing of water

The value of water has many social, religious, political,
and diplomatic overtones. However, the argument of this
book is that the most basic reason why inappropriate

7

habits of supplying and using water have persisted – with all the problems described above – is that it has been under-priced as an economic resource. Users do not, in general, treat water as an economic (that is, scarce) commodity, and the market is insufficiently used as a means of solving a problem of scarcity. This proposition is also true of its use as a receptacle of waste. Its capacity to assimilate and safely disperse waste has limits (i.e. is scarce). Pollution which is 'free' to a particular discharger can exceed that capacity and give rise to conflicts between different users, for instance industry, fishers, recreationists and households.

In some situations water is available to users at little or no cost, for instance, communities relying on natural springs from fully-recharging aquifers and farmers relying on seasonal flooding where the water has no alternative use. Many communities rely on small-scale roof catchments; once the relatively small installation cost has been paid for, this water is also free for all practical purposes. In another sense, extra water consumption has a very low cost: where investment in supply capacity is still under-utilised, the marginal cost of extra consumption is then little more than the energy cost of pumping the extra amounts. This is an argument for designing the tariff structure to encourage more consumption up to the point where capacity is taken up – but users could still be made to defray at least the average cost of the supply system, for instance through a fixed charge.

In the more common situation, however, water has a significant real cost of supply. Various kinds of cost are involved: environmental costs, to which we return shortly; the cost of provision; opportunity costs, consisting of the forfeit of the value of water in alternative uses; and in the case of finite sources such as groundwater, depletion. There are also costs, which we can ignore here, which fall directly on the users, e.g. household installations, on-farm pumping and land preparation; these all enter *private* calculations of the cost of water usage, but do not have to concern society as a whole.

Cost of provision

The cost of providing comprehensive public water services includes capital and recurrent outlays in supply, treatment and distribution, as well as drainage, sewage collection and treatment, and flood control, navigation and recreation. The principles involved in deriving prices from these costs have been well-rehearsed (e.g. Warford, 1968). The professional consensus is that tariffs should be based on the marginal cost of supply, interpreted as the cost of adjusting long-term capacity caused by a given change in demand (Munasinghe, 1992). The rationale for this pricing rule is that the use of water is optimised, in the economic sense, at the point where the benefit from the last (marginal) unit of supply equals the cost of providing that increment. If the benefit were less, society would gain by reducing supply to the point of equality. If the benefit were greater, there would be gains from expanding supply.

There are complications in following this formula:

1 where costs vary according to the time of use (e.g. domestic supplies, or seasonal farm use);
2 where costs vary between regions and over time. In Zimbabwe tariffs to farmers in the various provinces have varied from Z$9 per cubic metre to Z$278, depending on the cost of constructing the dam, which is correlated with its date.

Although such factors make pricing (especially the structure of tariffs) an imprecise art rather than a science, and complicate fine-tuning, they do not alter the need to fix water prices much closer to economic levels, even though there is room for dispute about what those levels should be. In most cases, the gap is currently so wide that any change which brings prices into the right general range is justifiable, even if there is room for debate over allowing for the complications.

Opportunity costs

There is an added difficulty where water has joint and/or conflicting uses and costs have to be allocated between the various users. Even in cases where the cost of water supply appears to be small, the water may have such an opportunity cost to society. Depending on circumstances, these might include power generation, urban or industrial consumption, the dilution of sewage effluent or farm run-off, as well as fishing, recreation, aesthetic pleasure, navigation, etc. These alternative values often change according to the time or season, for instance farmers value water more highly during the growing season, and it is more important to have high estuary water levels during low tides to flush out wastes.

If water had the characteristics of other goods, and were traded like them according to demand, these alternative values would be reflected in its price. However, because of the rarity of well-functioning water markets, its opportunity cost is rarely signalled to the consumer or polluter. Although it is impractical to incorporate opportunity cost into a standard pricing formula (because it varies between users and over time), the existence of alternative uses for water which are often more valuable than that of the target consumers reinforces the case for charging at least the economic price.

The above types of cost are recognised and are, in principle, measurable, but they are only captured in actual prices in a very minor way. Except where private managers or concessionaires operate (Coyaud, 1988), or in a relatively few well-organised utilities, prices for urban and industrial users usually fall well short of the costs of supply, narrowly construed. There are likewise few cases of effective pollution charges being levied (OECD, 1987; Bernstein, 1991). The situation in agriculture is even

10

worse. In practically every developing country (and most developed ones) the supply of irrigation water is heavily subsidised (Repetto, 1986; Postel, 1989). Water for rural domestic supply is also subsidised in most cases (Katko, 1990).

None of the above should be read as implying that there is no scope for reducing the costs of supply. Various examples are given in Chapter 3 of improvements to the distribution and retailing of water which can reduce waste, and therefore costs. The reform of water utilities often reduces costs. Staffing levels on many irrigation schemes in Africa are grossly excessive; where such schemes are turned over to farmers' own management, pumping time is economised and costs are reduced. But even allowing for the potential for such cost reductions, mankind still gets its water too cheaply.

Depletion

Aquifers which are drawn down at a rate faster than their natural recharge rate are akin to a mineral reserve. The Ogallala Aquifer underlying large areas of the Great Plains of the United States is one such reserve. Originally containing at least 100 years supply, it has been rapidly depleted, and now contains on average 40 years supply; some parts of it have been exhausted and the farms formerly dependent on it have been abandoned (Reisner, 1990). The huge scheme in Libya to extract and convey water from underground reservoirs in the desert to the coast could also be regarded as a mining venture. Many groundwater reserves underlying the large metropolises of developing countries are now being mined and contaminated. Principles used in the economics of mining can usefully be applied in such cases (see Chapter 4 for an exposition).

Water as a public good

Even if society wanted to recover the costs of supply, and charge polluters, the 'public-good' nature of the resource often makes this difficult. The essence of a public good is that it is available to all, and no one can be denied access to it. Thus a private agent has no incentive to invest in its preservation or improvement, since it would be impossible to recover costs from users ('free-riders'). Moreover, as the 'tragedy of the commons' (Hardin, 1968) highlights, no single agent has an incentive to refrain from exploiting it, since others would continue to do so; unless effective co-operation arises amongst users, there is likely to be over-exploitation and abuse of the resource (Magrath, 1989).

Certain water resources are public goods in this sense. A groundwater aquifer lying under a number of properties is one example; it is possible to control extraction by co-operative agreement, legislation governing sinking new wells, regulating extraction, adjusting the energy prices for pumping, etc. In practice however, this is difficult, and the widespread depletion of urban (and many rural) aquifers is proof of this. Unmetered piped water in the mains is also a public good, in the specific sense that the decision to consume more is costless to the individual user. Unless highly visible uses (e.g. car washing, lawn watering, swimming pools) are banned or controlled, in the absence of metering the water authority has no sanction on excessive and wasteful consumption.

Water as a receptacle for waste discharges is a clear case of a public good. In theory, wastewater discharge can be banned, regulated or charged for. In practice, most is not. In many countries the individual polluter has little or no incentive to refrain from pollution or to clean up the relevant water bodies. Hence they are polluted to the point where they become a public health risk, or impose costs on other users which are so large that they can no longer be ignored.

Unlike clean air or law and order – the textbook examples – water is not bound to remain a public good. It is capable of being brought under greater control, for the public benefit. This requires the development of public opinion and legislation, as well as the installation of administrative, policing and charging structures well beyond those now present in most countries. There are successful cases where such action has been taken, which offer encouragement to the rest (Bhatia *et al.*, 1993).

Environmental externalities

The use of water in households, industry, agriculture and other sectors incurs various kinds of environmental costs. Irrigated farming often returns water to rivers with a high saline and agro-chemical content, which is costly to households, other farmers, and fisherfolk. Releases of untreated industrial effluent can also poison fish, spoil rivers and lakes for recreation, and impose treatment costs on society. Diversions of water for power generation or irrigation reduce river flows, and can cause siltation, transport difficulties, dessication of wetlands, loss of fishing grounds, and the destruction of amenity.

These effects can, to some extent, be measured and subjected to economic valuation (Kneese 1984; Gibbons, 1986; Winpenny, 1991). Some costs fall on marketed output or money payments, e.g. loss of fishing income and tourist receipts. Others entail additional spending by public agencies or private individuals, e.g. extra treatment, cleaning, investment in new sources, etc. Other costs can be elicited by willingness-to-pay (WTP) surveys of how much users value environmental quality, or the travel costs they incur to visit an unpriced natural asset. It would be difficult, if not impossible, to capture all these costs directly in water prices. Some could be captured indirectly, via the recovery of the costs of compensation and damages from legal actions and criminal prosecutions brought by aggrieved parties. Pollution charges can also be levied at a rate that covers the cost of dealing with various kinds of discharge. In practice, they rarely cover more than this (OECD, 1989).

There is one important *positive* externality arising from the use of water, namely the social benefits to be gained from extending safe and adequate water supply and sanitation to the population at large. Consisting mainly of improved public health, but also including time and energy savings in collecting water, these benefits are usually regarded as social externalities; they accrue to society, and would not be fully captured in individuals' willingness to pay, because they would largely accrue to other people. There is undoubtedly some externality here, though a growing body of research indicates a high level of individual willingness to pay amongst the poor, as revealed in their transactions with private water vendors (Whittington, various; White, 1991). There is also empirical evidence of households' WTP for improved domestic sewerage (Darling *et al.*, 1992).

However, neither kind of WTP captures the social benefits of improved services. In practical terms, the existence of this social benefit means that schemes to provide such improvements can be credited with an environmental health benefit over and above the users' willingness to pay. And this should be reflected in pricing policies that encourage minimum levels of consumption, and safe sanitation (see the next section for indicators of the magnitude of this effect).

Despite these important external benefits from improved supplies, the average price charged for water needs to rise. There is no inconsistency here. The majority of users need to increase their contributions to provide funding both for its expansion to areas currently unserviced and from which cross-subsidies can be made to poorer users, if public policy so dictates. Poorer consumers can be encouraged to take minimum levels of water for essential household use by means of low, or even nil, charges for the first 'block' of consumption, with normal tariffs applying to subsequent increments. Well-managed water utilities can still pursue financial rectitude while providing socially desirable minimum consumption at special 'lifeline' rates.

CONSEQUENCES

Water is in universal use, yet it is consistently underpriced. In elementary economics, a commodity which is supplied too cheaply will sooner or later need to be rationed by more or less arbitrary means; those who are fortunate to be provided will use the water excessively and wastefully.

In most of the developing world, water services are far from universal and, in effect, are rationed. Despite the substantial achievements of the International Drinking Water Supply and Sanitation Decade (1981–90), over 1 billion people lack access to safe water, and 1.8 billion do not have proper sanitation (World Bank, 1992). And the backlog is rising in absolute terms.

The empirical evidence on the link between improved water supply/sanitation and health is suggestive of the costs of the present situation:

> inadequate sanitation and clean water provision remain, in terms of the scale of human suffering, the most serious of all environmental problems.
>
> (World Bank, 1992)

One exercise estimated the likely impact of universal adequate water supply and sanitation on the incidence of various diseases, as follows:

- 3 million fewer deaths due to diarrhoea each year among children under five (out of the current annual total average of 10 million deaths in this age group);
- 200 million fewer episodes of diarrhoeal diseases;
- 300 million fewer people with roundworm infection;
- 150 million fewer people with schistosomiasis;

- 300 million fewer people with trachoma, and 5 million fewer people blind from the disease;
- 2 million fewer people infected with guinea worm.

(World Bank, 1992)

There is, on the other hand, ample evidence of the excessive use of water by those with individual connections, or guaranteed access to water for commercial, industrial and agricultural purposes. In agriculture, which accounts for 80–90 per cent of the worldwide use of water, much irrigation is of low-value crops and there is enormous waste in its distribution and application. There are many systems where *most* of the water is lost to leakage or evaporation before it gets to the field (Postel, 1989):

> Only a small fraction of water diverted in most large surface systems in developing countries is available for plant use, typically 25 to 30 percent, compared to 60 to 70 percent in advanced systems . . . As a study of 11 major irrigation systems in China showed, for example, water use per hectare averaged twice and often exceeded three times the design application rate.
>
> (Repetto, 1986: 5)

In industry the extent of waste and misuse is evident in the large savings that are possible, some at little cost, by firms that are induced to make water economies as a result of pricing, rationing or other means. In the household sector, the gross under-supply of poor neighbourhoods coexists with high per capita levels of consumption among those with individual connections. In high-income areas within arid or semi-arid regions a large part of peak-time consumption is for 'frivolous' purposes such as watering lawns and swimming pools.

The low price discourages the technical change in water-using sectors that is desirable for the sake of long-term conservation of this increasingly scarce resource. Conservation by users (whether households, farmers or firms) is

16

not worthwhile so long as prices remain low. (Another motive for introducing conservation measures, especially by industry, is to safeguard against unreliable public supplies.) There are examples of major savings being achieved through industrial recycling which were only worthwhile to the firms themselves once an economic tariff for fresh water was fixed, or where pollution charges or other controls were introduced (see Chapter 3 for examples).

The failure to treat water as an economic (i.e. scarce) commodity has perverse dynamic effects, through its encouragement of a high rate of growth of water-dependent sectors. There is overexpansion of irrigated farming in a number of regions where water is actually very scarce (e.g. Sexton, 1990). Major water-using and polluting industries have been indulged through policies of protection and import substitution; the price of their water and pollution has not been large enough to influence their viability or growth. Underpricing urban household water can even be said to have encouraged the overexpansion of cities. Conurbations such as Mexico City, Santiago, Beijing, Delhi, etc. are starting to experience major water shortages, caused in part from the 'subsidy' to their expansion, due to the failure to charge residents and commercial users the full cost of their water.

In such cases a pattern of development has been launched which is unsustainable in the long run. The supply-led approach to water provision, coupled with a failure to price at an economic level, guarantees a long-term water 'problem'. It accelerates the rate at which readily available surface and renewable groundwater sources are allocated, and often leads to an irretrievable mining of the aquifer. It has been aptly remarked, à propos the Middle East, that the decision to bring water to a farming area makes a future shortage inevitable (Allan, 1992).

Much of the momentum for supply-led development is due to the way it succours the growth of powerful vested interests. Experience shows that when a scarce commodity is not allocated by the market mechanism, it tends to gravitate towards the rich, powerful or well-connected. The 'politicisation' of water allocation produces increasing diversion of public investment resources into supply schemes, and of recurrent budgetary revenue into subsidising the user sectors (including drainage, sewerage, subsidies for farm surpluses, debt servicing, etc). At the macroeconomic level, this can seriously distort growth (Sexton, 1990).

Low prices depress the profitability of investment in the water sector, and discourage private investment. In 1991 internal cash generation financed only 10 per cent of the costs of World Bank funded water projects, and the trend is getting worse (World Bank, 1992b). Privatisation has proceeded less far in water than in any other major public utility (Roth, 1987). The UK is a rare instance of the full private ownership of water assets. In many other cases, inspired by the French model, the private company manages the assets on behalf of the government, with its exposure to risk limited in various ways.

Much of the above also applies, *mutatis mutandis*, to water pollution. Pollution (e.g. industrial effluent, untreated household sewage, agricultural run-off and liquid animal waste) is tantamount to the excessive use of an unpriced public good, namely, the assimilative capacity of the environment. The failure to charge properly for this service encourages the growth of heavy water-polluting sectors, such as chemicals, metallurgy, food processing, pulp and paper, and intensive agriculture. Water pollution has costs of its own, but it also affects the supply of fresh water, where contamination of water sources occurs. These factors reinforce each other. A powerful vicious circle can develop, in which the long term scarcity of water is practically guaranteed. Pollution becomes very hard to control where large polluting sectors have been built up. Anti-pollution policies find it easier to deal with new entrants than with well-established polluters, many of which are large providers of income, tax revenue and jobs.

Under-pricing water is thus a major economic distortion. Research into comparative economic performance shows the advantages enjoyed by countries where a well-functioning market allocates resources to their most beneficial uses and helps to create wealth and growth (Agarwala, 1983). The fact that in many economies the low price of water is one of the most far-reaching distortions of all suggests the potential benefits from action in this area. A study of Jordan and Israel (Sexton, 1990) has highlighted the drag on overall economic performance from the failure to price water: an excessive public investment programme in the sector, with low rates of return, the promotion of uncompetitive user sectors, and the cost to public revenues from continuing subsidies to water use, and from disposing of farm surpluses. Other countries, such as the United States, could tell the same story.

Because of the large vested interests behind existing water pricing policies, the reaction of most policy-makers to suggested reforms would be to envisage political and economic *costs*. There would be powerful losers from reform in the short term – many consumers, the construction lobby, most farmers, etc. Emphasising and quantifying the *benefits* of reform – at a sectoral and macroeconomic level – help to place the debate in a more positive light, and to identify and mobilise the potential gainers (often, industry, tourism, high-tech farmers, environmentalists and less privileged consumers). We revert to the politics of reform in the final chapter.

The analogy with energy is suggestive of future trends in the water sector. Until the 1970s the low price of oil produced the same kind of distortions. The increases in oil prices during the 1970s set in motion structural changes leading to economies in energy use, and revealed an elasticity of demand for oil which few suspected at the time. The contemporary moves in the United States to mandate power utilities to conserve rather than sell power, is a

possible model for what could happen in the water sector
(Winpenny, 1992a).

2

TACKLING THE PROBLEM
Lines of approach

... in a barren and dry land where no water is.
(*Psalms*, l:xiii)

It was argued in Chapter 1 that the failure to treat water like other economic commodities lies at the heart of the problems noted. Because water is regarded as a special case, its planning has usually amounted to a search for the cheapest options of providing for projected requirements. This chapter reviews conventional planning and investment approaches, before introducing the philosophy of demand management, which underlies the remainder of the book.

The inhabitants of arid and semi-arid regions have always been aware of the value of scarce water. Laws and institutions have evolved to deal with its allocation, which have been more or less effective, though not necessarily efficient or equitable. Even in such regions, however, the scarce assimilative capacity of water has rarely been recognised, and consequently pollution is a growing problem almost everywhere.

For the larger number of people living in humid regions the realisation that water is not an unlimited resource, and that its receptor capacity is also becoming strained, has been very late to dawn. Laws, institutions and policies are confronted with a lot of necessary adjustment. Faced with evidence of future shortages of supply to meet growing demand, the typical response has been to: commission a comprehensive study of resources; project the demand on an 'unconstrained' scenario; consider the various supply-augmentation options; recommend that which meets projected demand at the least cost; and implement the scheme through public agencies, and at subsidised prices. The analogous approach to pollution has been to project future wastewater emissions and pollutant loadings, and to prescribe a combination of investment in treatment works and legislation to curb the most flagrant offences.

The keynotes of the traditional approach to water problems have thus been centralised planning and prescription, public agency, supply augmentation, subsidy, and reliance on administrative and legal instruments for allocation of supplies and pollution control. Measures to deal with these problems by managing demand have not been prominent in this stylised 'traditional' approach. In this chapter some of the shortcomings of existing approaches are discussed, and the merits of demand management through market processes are then restated.

CENTRALISED AND DECENTRALISED SOLUTIONS

An important distinction can be made between the appropriate unit for the *analysis* of problems, and that for *implementing solutions*. The two are likely to be different. In the water sector, the collection of data and the analysis of problems are best done centrally, while solutions are often best sought in a decentralised fashion, subject to the creation of the necessary legal, institutional and policy conditions.

Much planning starts from the view that the river basin is the 'natural' unit for water development since it forms a single hydrological entity, and the water cycle can be studied in a coherent way. While this may be true for study purposes, it does not necessarily follow that water development should be planned, and solutions sought, at this level. Many countries and regions adopt a larger planning frame. They refuse to allow the growth of certain areas to be limited by the hydrological capacity of the river basin in which they are situated, and have undertaken major inter-basin transfers. It sometimes makes sense to resort to international transfers of water (e.g. between Lesotho and South Africa) whether or not they are part of the same river basin. However, long-range transfers tend to be controversial in the regions of origin, and often have high environmental, as well as economic, costs.

The point is that geography and hydrology do not necessarily define the best scale for planning and problem-solving. Nor do they justify the use of 'integrated' or 'comprehensive' plans for water development, in which solutions are sought and implemented in a centralised manner. The use of such plans has been criticised for both practical and conceptual reasons.

Although professionals pay wistful tribute to the notion of integrated national water resource planning, there is little sign of it being used in practice in developing countries to tackle their more serious water problems. Progress in conserving and reallocating water depends in practice on solutions being sought and implemented at an urban or regional level – as the experience reviewed in Chapter 3 makes clear.

The impetus of integrated planning has been sustained largely by its intellectual appeal:

> Much intellectual effort has been invested over the past fifty years in the development of systems models which enable a small technical elite to better comprehend the complexity of man's interactions with the hydrological cycle. This 'systems thinking' has led to a powerful, appealing, intellectual paradigm of integrated water resources planning . . . . although an integrated view of water resources may be intellectually appealing, it does not necessarily follow that

water resources policies must themselves be comprehensive or well-coordinated to be effective.

In both developing and industrialized countries there is a huge gap between the concept of integrated water resources planning and how water resources planning and policy is actually done. In practice most water resources planning is done incrementally and is driven by the need to find solutions to relatively immediate, specific problems, not grand visions of river basin or regional development.

(Whittington, 1991)

Integrated planning, and its intellectual offspring the central allocation and management of water, in practice encourages rent-seeking behaviour on the part of its agents (Lovei and Whittington, 1991). Public utilities with a monopoly of piped water supply have the potential to generate sizeable monopoly rents, which may be diverted into the pockets of officials and meter-readers. The co-existence of public supplies to one portion of the population and the activities of private vendors servicing the rest creates a strong incentive for unscrupulous officers to limit public supplies and exaggerate their unreliability, while maximising (profitable) sales to the private vendors. This system is inefficient in that it increases the economic costs of distributing water. It is inequitable, in that the poor pay much higher unit prices for their water than those fortunate enough to have piped supplies. Rent-seeking also appears in the delivery of irrigation water, and even in the provision of sanitation. In Kumasi, Ghana, people pay each time they use a public latrine – quite literally 'spending a penny' – and half the revenue is appropriated by local political parties (Whittington, 1991).

Formally, as noted in Chapter 1, the optimal allocation system is that which equates the marginal cost of supply with the marginal benefit (shadow price) of the water in use. Neither a fully centralised nor a decentralised system is necessarily ideal on this criterion. Apart from incurring the risk of monopoly behaviour and bureaucratic inefficiency, centralised systems experience a high cost in collecting information, waste relevant signals about market behaviour, and discourage technical advance. Decentralised systems, on the other hand, cannot handle the extensive interdependencies and co-ordination problems that exist in a water system. In short:

> It is likely therefore that some mixture of top-down control, regulation and allocation of entitlements, bottom-up decision-making, and trading of water rights is efficient. Such mixed institutions have, in fact, been used with success, both with and without government sanction
>
> (Roumasset and Chakravorty, 1988)

Such a pragmatic stance is fully in keeping with the argument of this book, but will entail a great deal more stress on demand management in order to redress the supply-oriented bias in the traditional approaches.

SUPPLY AUGMENTATION

Many of the shortcomings of the supply-augmentation approach have already been alluded to, since they appear as the symptoms of water stress discussed in Chapter 1. To recap briefly, projects to increase supply are tending to encounter hydrological limits, face increasing costs in pumping or transferring water over long distances, entail increasing environmental costs to which the public is increasingly sensitive, and demand growing government subsidies. These are powerful arguments for shifting the current emphasis towards more careful management of the existing resource. The obvious objection to doing so is that many consumers do not have adequate, safe, supplies of water or sanitation, and the numbers of unserviced people are likely to grow with the growth of population and continuing urbanisation. How can supply to these groups be increased in a manner compatible with the principles recommended in this book?

In the first place, available supplies could be distributed more evenly and equitably, and finances should be made to reflect the new pattern of provision. In most existing systems available resources – the water itself, administrative and professional time, and financial subsidies – are concentrated on the more affluent consumers. Reducing subsidies, raising tariffs, and improving the rate of revenue collection would release funds for the expansion of the supply network. Wasteful and anti-social uses could be discouraged by progressive pricing and other, non-price, measures, releasing water for new users.

Secondly, waste and physical losses in the distribution system could be greatly reduced. The scope for mounting programmes for reducing unaccounted-for water is evident from the present size of losses from this source, 25–50 per cent being typical for many developing countries. But the profitability of such programmes depends on proper pricing.

Thirdly, there is great scope for encouraging the recycling of water in industry, the treatment and re-use of wastewater, and the introduction of efficient user appliances. These measures, which also depend on water being properly priced, enable a given source to go further.

Finally, as regards sanitation, there are a number of options for improving urban coverage with considerable cost savings compared to conventional systems. Apart from improved on-site devices (such as the Ventilated Improved Pit latrine) one option for cost-effective water-borne sewerage is effluent sewerage, which incorporates septic tanks between houses and mains sewers to enable smaller sewers to be laid at flatter gradients and with fewer manholes. Another promising system is the so-called 'condominial' pioneered in Recife, entailing a short grid of feeder sewers in backyards, and with the active involvement of householders in choosing the level of service and in maintaining the feeder sewers (World Bank, 1992).

In short, investment in supply augmentation along traditional lines is increasingly costly, in financial, economic and environmental terms. The imperative to invest in expanding supply and sanitation to growing populations and overcome the existing backlog will require major resources, but can be achieved only by reforming attitudes and policies in the sector as a whole. The case for a shift towards demand management is reinforced.

DEMAND MANAGEMENT

The case for greater reliance on the management of demand for water has to a large extent been made in the above criticisms of comprehensive planning and central allocation, on the one hand, and supply augmentation, on the other. Managing demand entails taking into account the value of water in relation to its cost of provision, and introducing measures which require consumers to relate their usage more closely to those costs. It entails treating water more like a commodity, as opposed to an automatic public service.

The aim of managing demand is to ensure that a given supply of water is distributed to accord more closely with its 'optimal' use pattern. In theory, this will be achieved when the marginal unit of water for each user has the same value (if this were not the case, total welfare could be increased by redistributing water). This theoretical ideal is unlikely to be achieved in the real world, but demand-management measures can help to move an existing allocation of water closer to that ideal. In practice, treating water as an economic resource means less waste, confining its use to where it is really valuable, and preferring reallocation to new supply schemes where these are costly in economic and/or environmental terms.

This book places emphasis on demand management not because these policies alone are sufficient for the water sector but because they have been neglected in the past. In any particular situation, both supply and demand measures will be required. Alfred Marshall likened supply and demand to the two blades of a pair of scissors. There will be occasions where supply schemes need to be developed, but it is important that all policies should be subjected to similar appraisal criteria. Chapter 4 discusses relevant performance criteria and appraisal methods valid for all types of projects and policies.

In 1989 the OECD's Council gave high-level endorsement to the role of demand management as an element in integrated policies for water resources. It recommended that:

> Member countries review their existing institutional arrangements, comprising administrative, legal and economic systems, in the field of water resources with the view to improving the integrated management of their water resource policies.

and that:

> Member countries develop and implement effective water demand management policies in all areas of water services through making greater use of: forecasting future demand for water; appropriate resource pricing for water services; appraisal, reassessment and transferability of water rights; various non-price demand management measures; and integrated administrative arrangements for demand management.

> (OECD, 1989: 182)

3

CONSTRUCTING THE POLICY MIX

> Dead water and dead sand
> contending for the upper hand . . .
> This is the death of water.
>> (T. S. Eliot, *Little Gidding*)

The argument so far has been that greater acceptance of market forces is necessary for an efficient and equitable solution to the future problems of the water sector. In fact, more is called for than just setting charges closer to economic levels – though that should be a centrepiece of any reform programme. Action is required at a number of levels to promote the more efficient use of water. This chapter classifies such measures as follows:

1 enabling conditions – action to change the institutional, legal and economic framework within which water is supplied and used (the 'rules of the game');
2 incentives – policies to influence the behaviour of users directly by providing them with an incentive to use the resource more carefully; these actions include both market-based and non-market devices;
3 direct interventions - through investment, spending programmes, or targeted programmes to encourage the use of water-efficient and water-saving implements.

These categories are set out in Table 3.1.

These three layers of policy are not alternatives, but

Table 3.1 Policy categories

Category	Actions and instruments
1 Enabling conditions	Institutional and legal changes Utility reforms Privatisation Macroeconomic and sectoral policy
2 Incentives	*Market-based:* Active use of water tariffs Pollution charges Groundwater markets Surface water markets Auctions Water banking
	Non-market: Restrictions Quotas, norms, licences Exhortations, public information
3 Direct interventions and programmes	Canal lining Leak detection Water-efficient user appliances Industrial recycling, re-use, water efficiency

strongly reinforce each other. For instance, the promotion of efficiency measures will be more successful in the context of the active use of tariffs. The promotion of water efficiency among household consumers is futile unless charges are sufficient to provide the necessary incentive. Water utilities have a better incentive to undertake programmes to reduce unaccounted-for water where the savings can generate adequate revenue.

Active charging policies, in turn, are more likely where water companies are privatised, or where their regulatory regime encourages them to be financially self-sufficient. Where efficient markets for water exist, higher prices mean that there are potential gainers (sellers) as well as losers (buyers). The establishment of markets for water will therefore create a group of supporters (e.g. farmers with water surpluses) as well as opponents. Some agricultural water buyers are potential sellers at other times, or even at the same time – where their holdings are fragmented.

Even if there is agreement on the broad direction and main ingredients of policy, the manner in which it is implemented remains very much an art. Is the best effect achieved with the simultaneous and full reform of all relevant policies and institutions (the Big Bang approach) or by gradual adjustments to accustom the public to the changed emphases? There is no general answer. An analogy can be detected with the debate over introducing market instruments in the economic reform programmes of former socialist countries. A number of interdependent actions are required to create the necessary critical mass and synergy for reform, but in some countries there is a case for gradual and progressive adjustment (see, e.g. Gelb and Gray, 1991).

There are also interesting issues in the optimal sequencing of reforms, for instance whether the price should be adjusted before or after reform of the water utility, or whether privatisation should precede or follow tariff reforms and improved revenue collection. As a general principle, governments can maximise their leverage over the sector by making sure that enablement and motivation are satisfactory. The extent to which governments become involved in direct intervention through spending programmes will depend on whether their enabling and motivating efforts need reinforcing, and on their financial and administrative resources.

The balance between the three layers of policy is also bound up with the mix of policy reforms as opposed to spending projects. Spending projects are often substitutes for necessary, but politically difficult, policy reforms. This is especially true when projects can be financed from foreign aid – e.g. subsidising efficient technology or appliances rather than raising the price of water. But even where a project approach is chosen, schemes to improve the efficiency of water use in industry or agriculture

depend on supportive macroeconomic and pricing policies, which may be a condition of the funding.

In practice, aid is available in various forms. External
support for structural and sectoral adjustment programmes means that finance is now available for both
policy reforms and projects. Governments are not con-
fined to implementing projects in designing programmes
to reform the water sector. It has even been argued that
the choice between policy-related and project aid can and
should be made on similar cost-benefit criteria (Kanbur,
1991). Policy reforms and projects can have strong mutual
benefits: policy reforms can increase the profitability of
projects, while projects that increase micro-level responsiveness improve the prospects for removing macro-level
distortions.

THE ENABLING ENVIRONMENT

Governments can set the groundrules for the supply and
use of water, within which utilities and consumers are able
– and motivated – to respond in a more 'rational' way.
The 'enabling environment' is a term that has arisen to
describe the creation of general conditions for encouraging
the more economically rational use of resources such as
water. It embraces various elements, such as institutional
and legal changes, the reform and privatisation of utilities, and setting appropriate general economic and sector-
level policies.

Institutional and legal reforms

Many of the problems being faced in the water sector can
be traced to the way in which both the supply and use
of water are planned, regulated, managed and financed.
The laws governing the use of water and the institutions
that have arisen to manage it are frequently obstacles to
making more rational use of the resource. In charting
reforms, however, there is no single successful blueprint:

Many types of institutions have been successful; indeed, there is no universally suitable model that can be prescribed. Institutions are the products of a country's history, society and economy. The choice of which institutions are developed is a local prerogative.

(Okun, 1991)

Amongst the many possible and successful models for institutional development in the water sector are the following: government administrative, regulatory and operating agencies at both national and local level; national and local quasi-governmental agencies; local public utilities; private companies owning and operating water utilities; publicly-owned agencies contracting with private firms for operation and management; river basin organisations, etc. (Okun, 1991).

Although the precise institutional form cannot be specified without some knowledge of local circumstances, a few general *desiderata* emerge from a review of experience. The rational use of water is commonly frustrated by the existence of numerous, overlapping and fragmented institutions in the sector. At the very least, responsible agencies should take a broader view and be prepared to co-ordinate their actions with those of others. One suggestion is:

The need is not for large new institutions but, rather, for small policy councils composed of cabinet rank officials who will be able to coordinate the work of the existing water institutions . . . [Policy] should emphasize structure, organization, management, and the role of adequate cost recovery.

(Rogers, 1990)

There is value in having an arm's-length relationship between the responsible government ministry and the organisation entrusted with water supply. This has a number of aspects. Firstly, the UK privatisation experience

33

showed the value in separating regulation and standard-setting, on the one hand, from water supply, on the other. Previously, water authorities had been both judge and advocate in their attempt to meet water quality standards.

Experience also shows that the organisation charged with water supply can easily be 'captured' by powerful users' interest groups. The subservience of the US Bureau of Reclamation to agricultural interests is well documented (Reisner, 1990). In Israel water supply has become inextricably linked with agricultural development, and the Water Commissioner is its servant (Sexton, 1990).

Water supply is often a natural monopoly at the local level, and for urban supplies this is usually the most efficient form of provision. But the market can be 'contestable' as a way of disciplining operators. Under this arrangement, competition is for, rather than in, the market, rather like the award of concessions for television broadcasting in the UK and elsewhere. Concessions and management contracts with private companies would consist of the award of national or local monopolies for a limited term, with renewal made conditional on performance. In Paris, for instance, the two major French water companies have simultaneous contracts to serve different parts of the city (Triche, 1991). In practice, contestability is usually associated with public divestiture, but it could also involve competition between public and private agencies, or between public institutions, for the right to supply services. Companies would understand that the award and renewal of contracts would depend on a minimum standard of performance and evidence of consumer satisfaction, and the presence of potential competitors waiting in the wings should be a spur to efficient and cost-effective service.

We should not make light of the difficulties of regulating private monopolies and setting performance standards that are realistic and monitorable. Regulation and supervision have to reconcile the need for public scrutiny and the application of sanctions for non-performance with the need to assure the operator of sufficient continuity and profits. This is a difficult balance to achieve.

Few would dispute the responsibility of governments to set and enforce environmental standards. This is a precondition of internalising environmental costs, for instance in water prices or pollution charges and fines. Enforcement is obviously important: several countries have model environmental standards on the statute books but show a poor record in enforcing them. The fact that many polluters are state enterprises or government departments makes it more, rather than less, difficult to enforce the law. In East European countries fines for pollution have been set at a low level, have not been collected in the majority of cases, and where they were collected were passed on as a cost of production and lost in inter-departmental transfers (Wilczynski, 1990).

The reallocation of water is often hamstrung by legal restrictions. Water markets can only develop where existing users have clear title to the water. In some countries or regions, users such as farmers risk forfeiting their rights if water is not put to 'beneficial use', which may not include its sale to others. In such cases, users have every incentive to continue their present usage, however unproductive. Likewise, for auctions to be possible the auctioneer must have uncluttered title to the resource, and the bidders must not have entrenched rights to it. Changes in water prices or amounts supplied are often governed by long-term contracts – in the US western states, typically 40–50 years for Bureau of Reclamation water. Contracts of this length frustrate the more active use of pricing.

Reform and privatisation of utilities

Certain of the benefits normally accompanying privatisation can be secured without a change in ownership. The previous section stressed the value of creating an arm's-length relationship between government and service company, which forces the latter to comply with economic and financial performance criteria.

Certain West African countries have applied the French model of the *contrat plan* to govern the relations between public utilities and government overseers. The utility has, in effect, to propose a corporate plan for approval by its sponsoring ministry, setting out its objectives, investment plans, pricing policy, etc., in a form which can be monitorable. Any objectives that are purely non-commercial, e.g. loss-making provision to deserving social groups, are clearly identified, so that separate financial provision can be made to cover them.

Most reforms of utilities have fiscal motives. Making public sector water authorities become financially more self-sufficient is an important reason for Hungary's forthcoming radical reforms. Starting in 1992, Hungary plans to take the responsibility for providing public water supplies away from the twelve regional authorities and transfer them to local authorities, together with ownership of existing assets. Existing water companies – which sell construction and operating services to the regional authorities – will be dissolved. State subsidies will, in principle, be abolished, except in hardship cases. Local authorities will be required to cover costs and raise investment finance largely from their own charges.

A major impetus for these reforms has been a desire to reduce the high level of state subsidies for water, which amounted to Forint 9 billion in 1989 ($150 million). In two years the level of subsidy has been reduced by one-third. As a result, charges have started to increase, quite dramatically in some areas, and sizeable differences in the level of charges are opening up between different localities. In two years the price of a cubic metre of water in Budapest has increased ten-fold, admittedly from a very low level. The level of charges in the Balaton region is five times higher than in Budapest.

There is no reason in principle why regulatory systems should not give private water companies incentives to maximise efficiency and conservation of use, rather than volumes supplied – in the same way that electric power companies and utilities are being given such incentives in the United States as a means of avoiding the economic and environmental costs of new investments. Although public utilities with some autonomy could also respond to the same regulatory pressures, private companies would have sharper financial incentives to do so.

There are degrees of privatisation, ranging from contracts for specific services, management contracts and leasing, up to full concessions – the latter involving responsibility for investment as well as operation (Coyaud, 1988). Although there are examples of public water authorities taking strong action over tariffs (for examples see Bhatia et al., 1993), other regimes find it politically expedient to delegate unpopular actions to private companies. In order to fulfil their mandate and make profits private companies, whether concessionaires or service operators, are forced to attend to pricing and efficiency questions. Good public utilities also do this, but the pressures on them are often to do the opposite.

Private companies already account for a high proportion of supply in several major economies. In France private water supply companies serve around two-thirds of the population, and 40 per cent of all sewerage services are private. In the USA investor-owned water companies account for 56 per cent of all systems. In the UK, even before the recent full privatisation of water supply, 25 per cent was supplied by private companies (Coyaud, 1988).

In the developing world, private water vending from mobile tankers is widespread. Private involvement in piped systems is more rare. French companies have become involved in, or have inspired, a number of countries, such as Morocco, Côte d'Ivoire, Guinea, Thailand, Malaysia and the French overseas departments. Regulated private companies also operate in Santiago de Chile and Guatemala City (Roth, 1987).

As the Côte d'Ivoire case demonstrates, however, the

institutional and regulatory environment of private companies can thwart the achievement of some potential benefits (Triche, 1990). Over the past 25 years the urban water sector in Côte d'Ivoire has been operated by a private company, SODECI (*Société de Distribution d'Eau de Côte d'Ivoire*) under a mixture of concessions and lease contracts. SODECI is a private company, established in 1960 as a subsidiary of SAUR, a large French water utility, to operate the water supply system of Abidjan under a concession contract. Subsequently the majority of the equity (52 per cent) was acquired by Ivorian shareholders and the shares are now traded on the Abidjan stock exchange.

For some years this arrangement performed well in important respects. By 1989 72 per cent of the urban population had access to safe water, compared to 30 per cent in 1974. In the rural areas 80 per cent of the population were served by water points equipped with handpumps, compared to 10 per cent in 1974 (though many of them were not in working order). There was a high level of operating efficiency in urban areas, with 12 per cent unaccounted-for water and a charge collection rate of 98 per cent for private consumers. Urban tariffs were high, almost certainly above the level of long-term marginal costs, especially for industrial users, in order to subsidise the rural programmes. Subsequent changes have been in a downward direction. Demand was depressed and revenues fell as industries recycled water – a desirable result provided it was not taken to uneconomic lengths – and used cheaper private sources. This was an important reason why consumption fell below projections, especially in more recent years.

Provision in the rural areas was less satisfactory. Rural investment decisions were made by the Water Directorate on the basis of over-optimistic consumption forecasts, and production of water was favoured over distribution. There was little co-ordination between the various players in the sector, responsibilities were fragmented, insufficient attention was devoted to maintenance, there was a lack of financial control and monitoring of SODECI, and the latter bore little risk and had little incentive to improve performance in the rural and sewerage sectors.

Privatisation in Côte d'Ivoire was effective in raising urban tariffs and curbing excessive consumption, especially by industry. The company also had a strong incentive to maintain an efficient urban system, with minimal levels of unaccounted-for water. It was much less effective in solving rural water provision. Despite cross-subsidies from the urban sector, rural systems were over-designed and badly operated and maintained, due largely to the terms of the contract between the government and the company and the absence of a proper division of responsibilities between the various agencies.

Several lessons may be drawn from this record (some of which were taken into account in the reforms started in 1987). Fragmenting responsibilities for planning, investment, operations, maintenance and debt service leads to a lack of accountability and inefficiency. Agencies lacked control over decisions affecting their efficiency. SODECI had insufficient stake in decisions affecting operations.

Rural and urban water supplies may call for very different financial and institutional approaches. Combining the two within the responsibility of a single operator can be problematic. Subsidising rural supplies out of a high urban and industrial tariff may be self-defeating if this reduces revenues and depletes the sector's financial resources. Large subsidies are also undesirable on wider efficiency grounds.

In the context of conservation, Côte d'Ivoire's experience shows that privatisation can be a good opportunity for raising urban and industrial tariffs, improving the unaccounted-for water ratio, and maximising revenue collections. In normal circumstances these are desirable, though in Côte d'Ivoire's case urban tariff increases may have been carried to unacceptable lengths because of the need to cross-subsidise rural from urban consumers.

To take another example, privatisation since 1985 in Macao has not brought an increase in the real value of the water tariff, but has led to increased collections, a doubling of the number of meters in use, and a continued reduction in system leakage to its present level of 11 per cent.

The above discussion has focused on domestic and urban water supply. Privatisation can also apply to irrigation schemes. Because irrigation accounts for a high proportion of all water use, the potential efficiency gains from reform can greatly outweigh anything possible in the urban sector. The reform of public irrigation systems has stimulated a large literature (e.g. Small and Carruthers, 1991). Turning irrigation schemes over to farmer management usually results in considerable savings of water, energy and other costs. If privatisation stimulates investment and the adoption of water-efficient technology, substantial water savings are possible. A study of California's San Joaquin Valley demonstrated that water use per acre was 4.17–3.69 acre-feet (af) with furrow irrigation, 3.57–3.18 af for shortened run (modified furrow) technique, 3.13–2.79 af for the sprinkler system, and 2.63–2.41 af for drip irrigation (Caswell et al., 1990). Water use in the most efficient (drip) system was only two-thirds that in normal furrow irrigation.

General economic and sectoral policies

Setting appropriate economic policies is a necessary, though not sufficient, condition for treating water as an economic resource. Starting from a situation of macro disequilibrium, the restoration of greater macroeconomic stability can have a number of benefits in the present context: it can reduce uncertainty, which, ceteris paribus, benefits long-term planning and decision-making and reduces the premium on immediate consumption and 'short termism' in policy-making. Macro stability also aids the efficient working of markets, which is necessary if water is to be dealt with as an economic good (Killick, 1991).

At a more disaggregated level, policies affecting import protection, export subsidies, taxes and subsidies on output and input prices, interest rates, and price fixing for key goods determine the incentives for the production and consumption of goods and services that differ widely in their 'water-intensity' and 'pollution-intensity'. All these economic policies can either support or frustrate the achievement of a more rational water system at the sectoral or user levels. The encouragement of more efficient irrigation practices and raising agricultural water prices will fail if crop prices strongly favour water-intensive crops and if other subsidies reinforce the prevailing cropping pattern and farm practices. In industry, whatever is done about industrial water prices would lose its force if protective policies favoured major water-using sectors.

In Jordan, for instance, expansion of the irrigated farm area has been a central objective of policy since the early 1950s, and has led to increased production of relatively low-value crops with high water demands, which are surplus to the local market and have been exported. This development is a serious economic distortion, with fiscal consequences, and is absorbing precious water in a water-scarce economy. National agricultural policy has confounded more rational water use (Sexton, 1990).

The promotion of water-intensive and water-polluting sectors like iron and steel, petrochemicals, and pulp and paper is deeply embedded in the industrialisation strategies of many countries. These industries typically pay only a fraction of the economic cost of their water, and little or nothing for their pollution. They often develop in an indulgent regime, with few incentives to curb waste or recycle water. Even if water and pollution charges could be raised to economic levels, their effect on water use would be buffered by an array of counter-signals: subsidies on other key inputs like power and raw materials; high protection against imports; ability to pass increases in costs back to the government or on to a monopoly state-owned customer ('soft budget constraints'); shortage of investment funds for water-efficient processes such as recycling; high import tariffs and/or overvalued exchange rates raising the cost of such equipment, etc.

For water pricing and pollution charges to be fully effective, there would have to be a radical change in the industrial regime in many countries. Reviewing industrial pollution in Turkey, Egypt, Yugoslavia and Algeria, Kosmo (1989) concluded:

> Although such measures as pollution charges, tax incentives, and subsidies are potentially useful, their economic significance is dwarfed by those of input and output pricing policy and trade and credit policy.

Water pricing measures would frequently have to contend with incentives to attract foreign investment in water-using and polluting industries, which create a countervailing policy environment. Certain countries have consciously sought to attract industries shunned by other countries for their environmental effects (Leonard, 1988).

The importance of potential changes in the structure of industry on the demand for water is evident in China. In Beijing and Tianjin six industries account for 85 per cent of all industrial water use. The water-intensity of production (measured in cubic metres per 10,000 yuan of output) is as follows in Tianjin: pulp and paper 994, petrochemicals 568, chemicals 314, food 175, textiles 104, and machinery 86 (Hufschmidt et al., 1987). A shift in the relative importance of these different sub-sectors – for example in response to changes in industrial strategy – would have dramatic effects on aggregate water demand.

MARKET INCENTIVES

The creation of an enabling environment needs to be complemented by creating specific incentives for the economically rational use of water. These can entail using the price of water to encourage its more efficient use, or can involve non-market devices relying on restrictions, *diktat* or persuasion. This section considers the first type of measure, relying on the creation of market-based incentives.

The market can be used in two related ways to promote a more economic use of water. Raising the price of water, or auctioning it to the highest bidder, is the most direct means of encouraging conservation and reallocation to higher value uses. Pollution charges based on the volume of wastewater are an indirect method of raising the cost of using water, with similar results for large users. The second approach is to raise the opportunity cost of using water by developing water markets. This will raise its real value and create an incentive for existing consumers to relate their use of water more closely to its marginal value, and sell the rest.

This section first considers pricing through water tariffs and pollution charges. It then discusses the various kinds of water markets, including auctions and water banks.

Water tariffs

As noted in Chapter 1, economic principles require that users be charged for water at a rate equal to the marginal cost of supply; at this level the benefit from consuming the last unit of water equals the cost of providing it. Applying this principle entails measuring the consumption of water (by metering) and making charges volumetric (proportional to the amount consumed). The assumption is that the marginal cost of supply (usually interpreted as the long run marginal cost, or LRMC) can be known reasonably accurately. Strictly, measurable environmental costs and benefits should also be included.

Although water tariffs are in widespread use in countries at all stages of development (OECD, 1987), they are usually perceived as a means of cost recovery rather than as a way of managing demand. Even those agencies that successfully recover the costs of the supply system do not *ipso facto* cover the LRMC, which is typically higher than average current costs. Metering, the precondition of volumetric charging, is far from widespread. Marginal cost pricing in conditions of increasing marginal cost (the most common situation) is best served by a progressive tariff structure, whereby higher units of consumption are charged at a higher rate. Many tariff structures do not, however, incorporate progressive pricing, and some display the opposite characteristic, especially for bulk users.

For reasons of equity, public health and amenity, there is a strong case for providing a minimum amount of water at low unit prices. Higher consumption should attract progressively higher unit charges, to reflect the increasing costs of providing capacity to meet peak load demand, and more generally to encourage consumers to save water and avoid frivolous use. These tariffs can then be fine-tuned to encourage off-peak consumption. However, progressive tariffs may not be appropriate where many households share a single water connection, or where poorer unserviced households buy water from others with a private connection.

The effective use of tariffs presumes that consumers will respond to higher prices of water by consuming less of it. In short, the price elasticity of demand for water must be significant, and greater than zero. There is substantial empirical evidence that this is the case (OECD, 1987; Gibbons, 1986). In developed countries (for instance, in Australia, Canada, the UK, Israel and the USA), it has been shown that the price elasticity of household demand for water tends to fall in the range −0.3 to −0.7 (i.e. demand falls by 3–7 per cent in response to a price increase of 10 per cent.) (Bhatia *et al.*, 1993). A smaller body of empirical evidence from developing countries broadly supports this finding:

> indoor residential demand may be more elastic than observed in US experience. Estimates in the range

-0.3 to -0.8 are usual. This is consistent with the greater importance of the cost of water in the household budget, and the relative acceptability of some substitutes for water use.

(Boland, 1991: 23)

There is growing empirical evidence of the responsiveness of demand to high unit water prices in poor metropolitan areas of developing countries (Whittington, various).

It should not be thought that reductions in consumption in response to the higher price of water will necessarily reduce its use below socially desirable levels. Consumption essential for personal, household and social purposes is usually a minor part of total use. In a typical modern household with piped water supply, out of a daily per capita consumption of 150–200 litres, only 3–6 litres would normally go on drinking and cooking, 15–20 on washing and personal hygiene, and 3–10 on cleaning the house. The largest household uses tend to be watering the garden, flushing toilets, and showers/baths (Bhatia *et al.*, 1993).

Industrial users also have the potential for dramatic responses to increases in the effective cost of water use in the form of effluent restrictions and pollution charges. In many OECD countries average industrial water use in the year 2000 is expected to be 50 per cent of what it was in 1975; in the United States it may be only one-third (Bhatia *et al.*, 1993).

Evidence of price elasticity in agriculture is more sparse, mainly because agricultural water prices are typically so low that they scarcely register as a significant cost in farm budgets. In many irrigation systems, the reliability of supply, the cost of other inputs, and output prices are far more important influences on water use. In many command areas, individual farmers have little or no control over the amount of water taken on their fields. However, activity in water markets, where the opportunity cost of water has a similar effect to a tariff, shows a good response from farmers, as we discuss in a later section of this chapter. For other purchased inputs, such as fertiliser, farmers show considerable price elasticity of demand, and it is reasonable to expect a similar response with respect to water charges.

In general, the traditional view that demand for water is price-inelastic is based on the historical legacy of very low water charges in many countries, which leads consumers to disregard water as a noticeable cost. Where water prices have been raised, and tariffs structured in a purposeful manner, demand has shown considerable elasticity.

The experience of Tucson, Arizona, has been closely studied in the water literature (Martin and Kulakowski, 1991; Martin *et al.*, 1988; Zamora *et al.*, 1981; El-Ashry and Gibbons, 1986). Tucson is located in a desert and was until recently almost wholly dependent on groundwater for its water supply. The advent of surface supplies from the Central Arizona Project will relieve the city's water position, but only postpone the need to reduce its total consumption, which is causing severe mining of the surrounding aquifer. Since the 1970s water charges have been periodically raised and adjusted to a two-part progressive structure that is a closer reflection of the real costs of supply. Pricing measures have been accompanied by conservation programmes. The combined effect of these policies has been a reduction in per capita water use. In 1980 water conservation became official policy of the state of Arizona, with a mandated goal of zero groundwater overdraft by 2025 (i.e. stabilising groundwater levels). Tucson has to set goals for per capita consumption, which, if exceeded, in theory attract a fine of $10,000 per day from the State Department of Water Resources.

Tucson's official policy of conservation has been steeled by the active use of prices to help manage demand. The rate structure adopted in 1977, and revised in 1980, was basically an average-cost system, but the incorporation of seasonal peak pricing and the increasing block structure (whereby successive increments of consumption attract higher unit rates) were steps in the direction of marginal cost pricing. Several econometric studies of water demand and its responsiveness to prices have come to the conclusion that the demand for water, though inelastic, is sufficiently responsive to price to make tariffs a crucial method of conservation. Exhortation ('preachment') is considered to have had an insignificant role in itself, except in making the public more aware of water prices. Per capita water use since 1976 has clearly remained lower than before the tariff changes of that year, despite a strong 'income effect' from rising living standards. Attempts to control for this and for climatic effects indicate a price elasticity of demand for water ranging between -0.27 and -0.70. The increasing block-rate tariff, although desirable in principle, is considered to have had little effect on demand, mainly because mean use is near the bottom of the schedule.

A number of features of the enabling environment may account for the qualified success of Tucson's polices. There has been a high level of public awareness and acceptance of the need for conservation. Given the city's desert setting, water is guaranteed a high profile, and is rarely out of the news. The fact that most of the water was drawn from a finite underground source that was clearly being depleted, helped persuade the public of the severity of the problem. The general principle that charges should be proportional to the cost of supplying different classes of consumer was accepted, and the main features of the tariff structure were intelligible in the light of that principle. A lifeline rate for small consumers (i.e. a low, subsidised, rate for minimum levels of use) was retained, and further helped to defuse resistance. With some exceptions, such as 1974 and 1976, prices have been raised annually by modest and uncontroversial amounts, and users have become accustomed to them. Tariff increases have been linked to the cost of providing an expanding network, dating back to the structure adopted in 1976. Now that water conservation has become the state's official policy Tucson is legally obliged to respond.

The city of Bogor, in West Java, Indonesia also adopted a more active water pricing policy. In anticipation of heavy increased demand on the system, the utility adopted a new tariff schedule and conservation programme in June 1988. An interim evaluation carried out shortly afterwards showed evidence of a positive consumer response (reviewed in detail in Bhatia et al., 1993). The new tariff increased the rate by between 100 per cent and 280 per cent, raised fixed service charges by 50–275 per cent, and increased the progressivity of the rate. A household with a monthly demand of 50 cubic metres would pay 140 per cent more than before. The tariff increase was bolstered by an information campaign.

Evidence from Indian and Chinese industry demonstrates an active demand response to water price increases. In two private Indian fertiliser companies of a similar size, the one paying a high price for its (municipal) water achieved a unit consumption 60 per cent below that in the other company, which depended partly on its own wells and partly on low-priced public supplies (Gupta and Bhatia, 1991).

In Egypt a clear link has been drawn between low water prices and the limited extent of recycling in the power sector and industry generally. More realistic prices for fresh water would make it worthwhile for industry to treat and recycle cooling water:

> Even though industrial water prices have risen tenfold in the past two years, they are still at most only 20 per cent of marginal costs . . . . the costs of treating cooling water may be economic for the power sector if water tariffs were increased . . . . the power sector accounts for 79 per cent of industrial water consumption.
>
> By reducing water usage and encouraging reuse of wastewater, higher water prices will also facilitate the

separation of toxic and non-toxic waste for treatment and safe disposal and thereby help to reduce water pollution. Industries such as chemicals and iron and steel (which are expected to increase their water consumption ten-fold by 2000 . . .) would have a greater incentive to conserve and reuse water. Presently, the rate of water reuse is only about 14 per cent.

(Kosmo, 1989)

One factor determining the elasticity of demand of household consumers is the margin of 'discretionary' water use (typically for outdoor purposes), or of wastage and leaks within the consumers' control. In the United States around half of typical household use in residential areas is for outdoor purposes (Gibbons, 1986). Industrial demand elasticity depends on the scope for reducing waste and adopting water-efficient and recycling measures, or using treated wastewater. In agriculture, elasticity is proportionate to the farmer's degree of choice over quantities grown, the choice of crops, the method of application, etc. Amongst poorer communities in developing countries, with low per capita levels of consumption and little discretionary consumption, there might nevertheless be substantial price elasticity, depending on the relationship between water charges and discretionary income. We revert to this issue in Chapter 4.

Price increases will only be effective in conserving water if the new unit price exceeds the marginal value of the water. In irrigation schemes, for instance, price increases are an uncertain method of producing conservation in areas where supply constraints are binding and prices do not 'bite' in farmers' decisions about the volume of water taken (Moore, 1991). In such cases, the value of the marginal unit of water can remain above its marginal cost. To have an effect on demand prices would need to be raised to at least equal the shadow price (marginal value) of water, which varies greatly between districts. Active pricing also requires strong political leadership to overcome consumer resistance.

In the Broadview Water District in the San Joaquin Valley of California, the authorities have introduced a policy of charging farmers significantly more for *marginal* units of irrigation water. Motivated by the need to reduce loads of boron, molybdenum and selenium in agricultural drainage water entering the San Joaquin River, the district discovered that the last 10 per cent of consumption was responsible for a disproportionate share of the volume of drainage, and thus pollution loadings. Farmers are set a level of approximately 90 per cent of average historical usage of water for each crop, to attract the normal unit price. Usage in excess of this is charged at significantly higher levels. The volume of drainage water collected in the district in 1990 was only 75 per cent of the average for 1986–8 (Keller *et al.*, 1992).

The impact of price increases on the poor can be mitigated through the tariff structure. 'Lifeline' rates are commonly applied to the first increment of consumption to avoid penalising poorer users and discouraging consumption considered desirable on social grounds. In considering the equity aspects of tariffs it is also relevant to note that adequate cost-recovery is a precondition of investment in new services and in maintaining the condition and reliability of the present network; on both counts the poor are vulnerable to shortages of funds in the water utility.

In agriculture, the effect of various water allocation and charging systems on equity has been tested using stylised farm models. Six systems were examined: an 'optimal' method under which an all-knowing authority distributed water in order to maximise total farm income or output; apportionment of water according to a farmer's share of irrigated land; volumetric pricing; acreage pricing; tax on output; and tax on purchased inputs. The second variant is by far the most common in use.

On the assumptions made, volumetric pricing produced

the same level of output as the 'optimal' allocation, and more than the other methods. Allowing for the empirical fact that the capital intensity of small farms tends to be greater than on large farms (and, by implication, their water use per acre is greater), it is found that the equity objective is best served by volumetric pricing, ahead of all other methods. Likewise for cost-recovery: volumetric pricing, depending on the level chosen, can be superior to other methods. In short:

> the current practice of distributing irrigation proportionally according to the size of holding, with virtually no or minimal charges for the supply of water in many developing countries such as India, Pakistan and Egypt, is neither economically efficient nor equitable, given the current economic realities.
>
> (Rhodes and Sampath, 1988: 116–17)

Pollution charges

The widespread acceptance of the 'polluter pays' principle has eased the way for introducing charges on effluent discharge. In principle, an economic charge would be related to the environmental damage caused by the discharge, or the cost of prevention, treatment or restitution, whichever is least. In practice, pollution charges tend to be set lower than this, to recover costs of monitoring, administration and, occasionally, treatment (OECD, 1987; Bernstein, 1991).

Charges for water pollution are of interest in the present discussion for their effect on the demand for water. Less pollution would also safeguard more water sources for consumption purposes, and reduce the cost of treating contaminated supplies. In discussing the appropriate control regime for industrial wastewater, we need to distinguish consumptive from non-consumptive water use, and the quality from the quantity of wastewater discharged.

Industries with a high consumptive use effectively

extract water from the system and deny its use to others. This entails a higher net cost of supply compared to that of a user with the same water intake, but a low consumptive use, where a high proportion of the intake is returned to the system as waste. In the latter case, the wastewater is available to other users (ignoring, for the moment, its quality) and serves the useful environmental purpose of diluting waste already present in receiving water bodies and rivers. This picture is complicated where wastewater is returned in locations that are inconvenient for other users, in which case it may be effectively taken out of the system – e.g. if it is discharged into the sea, or into another river system where dilution of waste is less important.

In principle, industrial users should receive credits or rebates for the amount of wastewater they discharge, provided it meets quality standards. This does not affect the argument for charging economic prices for their water intake, which, in the case of high consumptive users, would stand as an incentive to restrict water use. A model industrial water regime would include economic prices for supply, penalties for wastewater pollution content, and credits/subsidies for the volume of wastewater of acceptable quality (Sunman, 1992).

The environmental impact of industrial effluent depends on the amount and type of pollutant it contains, whether or not toxic substances are present, its temperature and colour, the location of the discharge, as well as its volume and variations in the latter. Hence pollution charges are often combined with 'command and control' regulations to take account of these local factors. In the UK consents to pollute are issued for payments graduated to take account of such local factors and the expected amount and type of pollution (NRA, 1991).

Brazil's experience with the effect of pollution charges on water use shows substantial price elasticity. In three industries in Sao Paulo the introduction of an effluent charge led within two years to a 40–60 per cent reduction in water consumption (Miglino, 1984). In the Netherlands industrial water consumption fell by 30 per cent in the period 1970–76 following the introduction of water pollution charges in 1969, at a time when industrial output increased (quoted in OECD, 1987).

Pollution charges are effective in inducing firms to reduce their demand for fresh water, and are particularly useful where it is not feasible to charge firms for using their own water sources, such as wells. Provided certain minimum standards of water quality can be met (e.g. by refusing permits for toxic discharges or outlets close to areas of public use) volumetric charges can serve environmental purposes as well as providing an incentive for efficient water savings. There are no clear-cut equity concerns: application of the 'polluter pays' principle will benefit poorer groups insofar as they are affected by environmental pollution.

Other measures include subsidies, tariff concessions and tax incentives for the use of in-plant treatment and equipment for recycling, which can be used independently or as a 'stick and carrot' approach coupled with pollution charges. Measures to encourage the re-use of water also reduce the demand for fresh supplies, e.g. treating industrial wastewater or household sewage to a standard sufficient for its application to agriculture, or persuading industry to treat sewage for its own use.

Whether recycling is an economically efficient method of conservation, compared with alternatives, obviously varies according to circumstances. Recycling both cooling water and wastewater were superior to developing new supplies in Beijing. In two large plants in the Madras region recycling, condensate recovery, the use of regenerated water and other measures enabled firms to double their operating capacity without increasing their demand for water. In Jamshedpur, however, re-using treated industrial effluent was uneconomic (the evidence is reviewed in Bhatia *et al.*, 1993).

Water markets: general aspects

Where users have entrenched rights to water supply, real-location is only possible if they can be encouraged to sell some of their water to others, presumably for higher-value purposes. The existence of alternative outlets for water creates an opportunity cost for its continued use by, say, a farmer. All water, not just that which is 'surplus' to the farmer's use, becomes potentially marketable, and farmers have an incentive to drop low-value applications if they can earn more by selling the water. Moreover, the economic and environmental costs of developing new sources would be avoided.

In some regions agricultural users have no legal rights to the continued supply of water, and in such cases reallocation is within the discretion of the water authority, with or without compensation for the losers. In parts of Gujerat, India, farmers on public irrigation schemes have no entrenched rights to water, and the authorities are diverting supplies to urban use. Direct action of this kind is inferior to market solutions in that there is no presumption that all the water transferred goes to higher value uses, and there is no built-in compensation for the losers.

There are examples of inter-firm water markets in Indian industry. Unlike in the power sector, where companies with power to spare from their own generating capacity are not allowed to sell either to the national grid or to other firms, there is no restriction on firms trading water. Large companies with their own captive supplies or with a surplus from their recycling or treatment processes can and do sell water to other firms, some of whom are too small for water-efficiency investments or the treatment and re-use of effluent to be economic.

Groundwater markets

The markets for groundwater in Gujerat, India, have existed for 70–80 years, and they are also well developed in Punjab, Uttar Pradesh, Bangladesh and elsewhere in Asia (Shah, 1985; 1989). Owners of wells to all intents and purposes have ownership rights over the water they draw, and sell surplus water to other farmers. Although the typical transaction is on a temporary basis, there are a number of large-scale water dealers selling large quantities to regular buyers through a highly capitalised network of pipes.

It is common for farmers to be both buyers and sellers of water at different times or even at the same time in different locations. From a static viewpoint, the existence of the groundwater markets contributes to the efficiency of water use: water tends to find its way into the hands of those farmers who can use it to its highest value, and the knowledge that water is available for purchase discourages farmers from investing in their own wells at uneconomic cost. The distributional effects of these markets are not clear-cut. Large farmers predominate as sellers, small farmers as buyers; but the latter avoid having to make their own investments, and prefer to buy in rather than depend on unreliable public surface supplies.

The major problem with groundwater markets is the environmental effect of aquifer depletion. The existence of profitable outlets for water encourages greater pumping. If aggregate pumping exceeds the recharge rate of the aquifer, water 'mining' occurs, with environmental costs. There are signs of this happening in parts of Gujerat. This is a classic case of 'market failure' due to excessive use of a common property resource – the underground aquifer.

Water auctions

Water auctions are rare. The evidence considered here is taken from Victoria State, Australia (Simon and Anderson, 1990), and from Alicante, Spain (Reidinger, 1992). A precondition of the auction is that the authorities have a free hand in disposing of the water to the highest bidder, which implies that consumers do not have customary or legal entitlements to the water in question. The auction enables users to reveal their valuation of the water and the public supplier to extract the surplus (or rent) from the sale. If market prices are a reasonable echo of economic benefits, an efficient auction would also maximise the social benefit from using the water. Potentially, an auction could score highly on economic efficiency grounds, provided there is no collusive or monopsony behaviour on the part of the bidders.

In Victoria six water auctions were held in 1988 and 1989 in which the Rural Water Commission disposed of 30,000 megalitres of new irrigation water to farmers. The auction was controversial, and generated much local hostility. The total amounts were small in relation to normal requirements in the northern part of the state (c. 2.5 million megalitres supplied to 18,000 farms) and the successful bidders were those seeking marginal supplies for insurance purposes or for high-value applications. Dairying, fruit and livestock production are typical activities.

The Victoria example illustrates the problems of getting the reserve price 'right'. Quantities of water remained unsold, implying that the reserve price was above the users' valuation. It also revealed the existence of a trade-off between efficiency and equity. The auction was designed to protect the position of the smaller farmer, and improve its public acceptability, but in so doing it limited its scope and prevented maximisation of bid prices and revenues. The extent to which reallocated water was actually applied to high-value purposes was quite limited, partly due to the above factors, and partly to the purchase of auction water for drought security rather than use on new high-value crops.

The auction in Alicante is a long-established institution, integral to the operation of the *Huerta* (irrigation service area) of Alicante, which dates back hundreds of years. The irrigation area comprises 3,700 ha of good land, receiving water from a reservoir (owned by the water users' association), two inter-basin transfer canals and local groundwater. Every Sunday morning a public water auction is held in San Juan, a pueblo in the middle of the Huerta. The market is for tickets representing the right to take a fixed flow of water for a certain amount of time during a cycle of canal flows. Only enough tickets are sold to correspond to water time actually available.

The system appears to work well in allocating available water to the highest value uses during the season. There is a strong demand for scarce water. Total supplies are less than one quarter of what the area could use. Members of the water users' association that owns and operates the irrigation scheme own or control only 30 per cent of the water they need – the rest being bought. Water is paid for and delivered by volume – periods of flow at a constant flow rate. No attempt appears to be made by government to influence the going price or to favour certain users over others, and third-party or environmental effects do not seem to be prominent.

Surface water markets

Most of the evidence on the evolution of surface water markets has been derived from the Western states of the USA. (See Saliba and Bush, 1987; Frederick, 1986; Robinson and MacDonnell, 1990; Young, 1986; among others). Within the United States, states differ in the degree to which users may legally transfer water amongst themselves and to other parties. The Bureau of Reclamation, after a period of antagonism, is now relatively permissive in its attitude to the resale of its own water. It authorises such transfer provided third parties and the environment are not harmed, and that the transfers do not harm the federal government financially, operationally, or contractually (Moore, 1991).

A precondition for water markets is the existence of enforceable property rights to the water. This entails basic legal infrastructure, courts, and potentially lengthy and expensive adjudication – examples of 'transaction costs'.

Other transaction costs arise from the trouble of putting buyers and sellers into contact with each other (although these services are not slow to develop as markets grow) and of safeguarding the position of third parties and the public interest. There are also the physical costs of transfer and storage. For water markets to flourish, the efficiency gains from transferring water to higher value uses need to be large enough to offset the transaction costs.

There is little doubt that water markets in the Western US function well in allocating water to higher-value uses within agriculture, and between agriculture and urban/industrial sectors. The process is efficient and – at least between the parties to the transaction – equitable, since sellers are compensated at market prices. Whether it is socially optimal depends crucially on whether third-party and environmental effects are internalised in some way.

In the cases considered in one recent assessment (Saliba and Bush, 1987), third-party interests, including downstream users whose entitlements would be affected by increased upstream diversions, *were* largely incorporated into market decisions (though with some transaction costs). This conclusion appears to refer to third parties with clear and enforceable rights to water, who can stake a claim for compensation. Third parties with a more distant interest are more difficult to compensate, especially where they are in a different state (Young, 1986). There are also effects on the neighbourhood and region from diminished water supplies, offset by gains to the area receiving the waters. These effects are more difficult to evaluate (Whittlesey, 1990). The environmental effects of water market transactions are often left out of account in the values that are struck between the parties. The impact on in-stream values (fishing, recreation), water quality, wildlife habitats, amenity etc. is felt by parties who are usually 'underrepresented in the market place' (Saliba and Bush, 1987).

The basic preconditions for the further development of water markets are legal and physical. Sellers must have clear legal title to their water and the freedom to sell. Likewise there must be the physical means to make transfers feasible and economic. Given these conditions, the key to the wider development of markets clearly lies in the adequate recognition of the various third-party and environmental effects. This will, however, set up a trade-off with efficient private exchanges:

> Public policy must seek a balance between unrestricted markets which can impose high external costs, and market restrictions which reduce external costs, but make transfers more expensive, both to market participants and to agencies which evaluate transfer proposals.
>
> (Saliba and Bush, 1987: 256).

One of the longest established and most successful water markets is in the Colorado-Big Thompson (CBT) scheme (Howe *et al.*, 1986). The CBT transfers water from the western slopes of the Rocky Mountains to north-eastern Colorado, where its distribution is the responsibility of the Northern Colorado Water Conservation District (NCWCD). Since 1957 the CBT has provided an average of 230,000 acre-feet, or 17 per cent of the total water supply of the region. Although CBT water is used mainly for supplemental irrigation, it is increasingly also used as a fresh water supply by urban and non-agricultural industrial consumers. The markets that have evolved are unusually efficient, and might serve as models for study by other systems.

Since the mid-1960s urban and industrial growth has been rapid on the eastern flanks of the Rockies. Most of the water needed by these new sectors has been provided by the transfer of NCWCD allotments from agriculture. In 1957 irrigators started with 85 per cent of the allotments, whereas by 1982 their share had fallen to 64 per cent (though this exaggerates the shift, since cities tend to acquire more allotments than they are likely to need, and rent back the surplus to farmers). Average allotment prices increased by a compound annual rate of 19 per cent from 1960 to 1973, and by 33 per cent to 1980, before falling back in 1985 as a rival source of water came into being.

CBT water delivered through NCWCD auspices is in allotments that are uniform, easily transferred and reliable – all of which assists market creation:

> the NCWCD system is more efficient than the typical Bureau of Reclamation contractual arrangements, which tie water perpetually to the same land and, in many cases, to the same uses . . . . Inflexibility in patterns of water use like those found in central Arizona and the Central Valley of California either stifle further economic development or require enormously expensive new water projects to supply water for growth. The Central Arizona Project and the California State Water Project are two of the most expensive projects ever undertaken anywhere in the world. The well-known study of the role of water in affecting the growth of the Arizona economy . . . showed decisively that an efficient transfer of relatively small amounts of water out of low value agriculture to the newly-emerging urban and industrial uses was adequate to maintain rapid state growth without the Central Arizona Project. In a water market system, such transfers take place.
>
> (Howe *et al.*, 1986)

The success of the CBT water markets was helped by the fact that its water was widely held throughout the area. Since it had the lowest transaction costs, it represented the easily tradeable margin. Moreover, return-flow externalities were not an issue in this case, since the district owned return flows. The development of the market for allotments was supported by a majority of users from the outset. In addition, making more efficient use of existing supplies avoided the costs of developing new supply sources, including compensating the basins of origin, and avoided the increasingly high cost of conflict resolution in Western water issues.

Industrial water markets and exchanges

There is evidence of industrial companies buying and selling water in situations where public supplies are scarce and/or unreliable. Major industrial users may opt to develop their own supplies (e.g. from groundwater, surface diversions, treating effluent or sewage) or take other steps to secure their future supplies. Certain firms may have surpluses of water in the short term or even permanently. Companies differ in their access to supplies and in the cost of securing them. It would therefore be unsurprising if markets in water developed in industrial areas analogous to those in certain agricultural regions.

In Jamshedpur, India, TISCO, an integrated steel plant owned by the Tata Co., sells surplus water to other Tata firms in the area. A Birla subsidiary in the city buys water from tankers, which comes indirectly from Tata supplies. In the Madras region all firms face growing water shortages, and the larger ones have started introducing recycling measures and the tertiary treatment of municipal sewage. These measures are expensive, and some have a high minimum economic scale, which makes them prohibitive for smaller firms. Faced with these large differences in the cost of securing water, there is great potential for trading between firms (Bhatia et al., 1993).

The extent to which industrial water markets actually develop will depend on firms' relative costs, security of access to their own or public supplies, the physical transferability of water, competitive factors, the ability to make enforceable long-term contracts, etc. There is a close analogy with the power sector, where firms have their own captive supplies and can, in theory, sell surpluses to the grid or to each other. Where spare capacity already exists, it will usually be efficient to use it, as opposed to investing in new public or private supplies. Where capacity has not yet been created, its efficiency depends on the result of its cost-benefit analysis, compared to other options including expanding public supply (see Chapter 4).

Industrial water markets would be equitable if smaller firms were to pay less for reliable supplies than if they use other sources. Domestic consumers might also benefit from reduced 'crowding-out' by industry. A major drawback would be the environmental cost if there was encouragement for the increased drawdown of captive groundwater sources.

Water banks

Banking is an elementary method of storing water at a time when it is not needed in order that it can be drawn upon later, or in the meantime by someone whose need is greater. The simplest kind of banking is allowing unwanted surface water to replenish the underground aquifer, where it is available to be pumped in future. Other types of banks, discussed below, allow other parties to use the water on a temporary or permanent basis.

A water bank was created in California in February 1991 in response to the recent drought. The Department of Water Resources (DWR) bought 750,000 acre-feet of water from farmers on a 'willing buyer, willing seller' basis to guarantee critical supplies for urban and other purposes. The water was made available by means of putting farmland under fallow, using groundwater rather than surface sources, and drawing on reservoirs. The Water Bank paid sellers a price of $125 per acre-foot (af), based on an analysis of farm budgets and discussions with other interested parties. By the end of the 1991 season, the Bank had bought 831,000 af, 50 per cent from fallowing, 33 per cent from groundwater, and 17 per cent from storage (Keller *et al.*, 1992; Kennedy, 1991; Vaux, 1991; DWR, 1991).

The selling price was fixed at $175 per af, plus specific conveyance costs. Buyers had to establish that they had 'critical need', that they were using available water properly and carefully, and that they had implemented a water conservation programme. Of the total amount purchased, the Bank delivered 391,000 af (47 per cent), and retained 272,000 af (33 per cent) for storage. The remainder (20 per cent) was lost in conveyance and storage.

The impetus for forming the bank was four years of drought which caused reservoirs to fall to very low levels. In February 1991 deliveries to agriculture from the State Water Project ceased, and urban deliveries fell to 10 per cent of normal levels. The Governor was under pressure to appropriate and redistribute Sacramento irrigation water. Thus the situation was an emergency, with a high level of public concern, and with the alternatives involving high political costs:

> The Water Bank is perceived as an effective, short-term emergency program that was developed and executed extraordinarily quickly. It generated a broad range of support and was a useful learning experience, but it is not clear to what extent, or in what form, it will be institutionalized.
>
> (Keller *et al.*, 1992: 14)

Given the haste, and the unusual circumstances, certain features of the bank were accepted which would need to be scrutinised more closely if this were to become a permanent feature, or to be replicated elsewhere. One is the basis for price-setting. Prices offered to farmers, although ostensibly on a 'willing buyer, willing seller' basis, were set artificially in the absence of a normal market. Although the DWR was in a monopsony (sole buyer) position, it was also paying a scarcity value for the water, and its prices were considered high. The emergence of a normal water market in this region is bedevilled by the hydrological complexity of the Central Valley stream system and the limited conveyance capacity. The Department was mandated to take account of effects on fish and wildlife, though in practice transporting the water proved difficult to reconcile with fish migration schedules. It is also feared that the attractive prices offered for groundwater have encouraged localised overpumping of the aquifer.

The water bank seems to have been successful as an emergency measure and it has been acceptable to the public. As a longer-term solution, however, it is less efficient than a full private water market might be, though the physical problems in setting up such a market should not be overlooked. It has not fully come to terms with third party and environmental costs, and it entails a high degree of public intervention.

California is the setting for another water banking case, arranged between the Metropolitan Water District (MWD) and the Arvin-Edison Water Storage District (AE) in Kern County in the south-eastern part of the San Joaquin Valley. AE contains 113,000 acres of irrigated farmland, 46 per cent of which is supplied with surface water from the Central Valley Project of the Bureau of Reclamation and the California Aqueduct. The remainder is irrigated from pumped groundwater.

When available surface water exceeds the demand for it, the AE delivers it to spreading grounds where it is allowed to percolate into the aquifer. When surface supplies are insufficient, this groundwater is pumped up. Since 1986 AE and MWD have been discussing a joint storage and exchange programme designed to enable both districts to stabilise their water supplies. In wet periods the MWD would deliver water to be stored in AE's aquifer. When the MWD needed water, the AE would pump the aquifer for its own needs, and allow MWD to keep water from the California Aqueduct that would otherwise go to AE.

The agreement should enable both districts to make more efficient use of infrastructure and management, without affecting other water users. When water is delivered during the wet season, the Canal has excess capacity. In the dry season AE will be pumping, and reducing the amount of water subject to a summer surcharge. There should also be environmental benefits insofar as there will be less pumping of water from the Sacramento/San Joaquin Delta during the dry season (Keller *et al.*, 1992).

Transferable water-use permit

The transferable water right, or use permit, has existed in the Murray Basin in New South Wales, Australia since 1984. Surface water allotments for irrigation, industrial, recreational or environmental purposes have been tradeable, subject only to the veto of the state on water quality grounds. Transfers of groundwater can also be made. By 1988 surface transfers accounted for 5 per cent of the total surface water use in the state (OECD, 1989).

The transferable use permit has also been proposed as a method of persuading farmers in the Western states of the USA, currently with long-term contracts for Federal Bureau of Reclamation water, to conserve and reallocate their supplies (Moore, 1991). The Bureau is the major supplier of irrigation water from surface sources to Western farmers. In 1987 it serviced 10 million acres of cropland, one-quarter of the total irrigated area. The prices embodied in its long-term contracts with farmers are invariably well below economic levels. The Bureau has always had the freedom to excuse its customers from paying the interest element in its supply costs, and it has made extensive use of its ability to charge according to 'ability to pay'.

The Bureau has now been charged with responsibility for conserving water. This responds to three growing pressures – the need to reduce usage by irrigated farming in order to release supplies for growing urban consumption, to maintain river flows for in-stream environmental benefits (e.g. fish and wildlife, recreation, dilution of pollution), and to meet obligations to Native Americans.

Amongst the options for promoting conservation, quantity restrictions would be an option only when contracts came to be renewed. This option would be simple to administer, certain in its effects, and would leave irrigators free to adjust to it in their own profit-maximising way. It would, however, leave the economic rent from subsidised water with the farmer.

The price of irrigation water is normally laid down in long-term (typically 40-year) contracts. Increases could legally be made only when contracts came up for renewal. Confining price increases to new or renewed contracts would be fair to farmers for whom the irrigation subsidy has been capitalised in the price paid for their land. However, raising prices would not achieve conservation if the resulting price were still below the marginal value of the water (or shadow price), because it would still be worth the farmers' while to use the same volume of water as before.

Research in 18 irrigation districts in various Western states revealed that in six districts the contract price equalled the shadow price, implying that any price increase would produce some reduction in water use. In the remaining 12 districts, the shadow price of water exceeded its contract price, implying that irrigators faced a quantity limit on the amount supplied, and would use more of it if they could. In these cases, the price would have to be raised to the shadow price to have any impact on conservation. In those cases (4 in this sample) where the full-cost price was below the shadow price, cost-recovery alone would not produce the desired response.

Price increases would, however, have useful fiscal benefits, and would extract some of the rent conferred by irrigation water subsidies. Under a scheme of transferable water use permits, existing contract-holders would be allowed to sell their water rights as if they were private property. This provision could be applied to all existing contracts, provided it were consistent with state and inter-state water law.

As with water markets and banks, the permit scheme would not of itself acknowledge third-party and environmental concerns. It has therefore been suggested that a hybrid programme might be the most effective option, combining quantitative restrictions (to satisfy environmental needs and Native American claims) with transferable permits, as in federal programmes to control air pollution and lead emissions. This option would achieve conservation aims with certainty, would provide offsetting benefits to farmers wishing to trade, and would have no fiscal implications for government. In Australia similar objectives are achieved by allowing the state government to veto exchanges on environmental grounds. Alternatively, the state government could acquire the right to outbid offending transfers in the public interest.

NON-MARKET INDUCEMENTS

Non-market devices take a variety of forms, such as laws and sanctions, administrative *diktat*, persuasion and example, public education, etc. The most basic distinction is between compulsion and persuasion.

Restrictions and legal sanctions

Users can be compelled to conserve or reallocate their water by various means. In an authoritarian system, where consumers have little power, water can be turned on and off, and reallocated at the discretion of the system managers. Supplies can be cut off at times of shortages, causing involuntary conservation. Such measures are effective (e.g. in Gujerat for diverting water from agriculture to urban sectors, and cutting off irrigation water to Californian farmers in 1991) but do not guarantee either efficiency or equity.

In other cases, legal sanctions are applied to users who offend against behaviourial norms laid down by law. Restrictions are commonly placed on certain activities as a response to temporary or seasonal drought and shortage (e.g. bans on the use of outdoor sprinkling or hosepipes in the UK and Perth, Australia, (Hanke, 1982)). Rough notions of efficiency and equity can be served by targeting non-essential and low priority applications, and such measures can reduce consumption, even over an extended period (as in Perth). But their success depends on a high level of public compliance, which relies on widespread understanding of the problem and support for the privations that are caused.

Quotas and norms

Quotas and norms may be set for water users as an attempt to allocate scarce supplies in an equitable manner. The 'rationing' can be effected by issuing fixed quotas and monitoring their compliance, or by charging penal tariff rates on levels of consumption exceeding the norms. In the two cases quoted below good results appear to have been achieved by combining norms with penal pricing for those who exceed them. Israel and China have used hybrids of 'command and control' and economic instruments. Penal tariffs have a psychological effect similar to a fine, but they are more efficient since the charge is in proportion to the 'excess' consumption, and firms that badly need the extra water may continue to draw it. It may be no coincidence that both these countries have an extensive state apparatus and a citizenry that can be highly motivated to serve particular national objectives.

Israel has a comprehensive system of industrial water licensing, based on norms which take into account best-practice technology, updated from time to time and modified by the specific circumstances of each firm. Firms exceeding these norms are levied a penal surcharge of 200 per cent of the standard rate. Between 1962 and 1982 average water consumption in Israeli industry, expressed per unit value of output, fell by 70 per cent (Arlosoroff, 1985).

In Tianjin, China, norms are promulgated for industrial consumers based on regular detailed water audits, and users who exceed their norms attract a penal rate of up to 50 times the normal charge, depending on the extent of the transgression. Between 1981 and 1988 Tianjin achieved a 250 per cent increase in the value of output per unit of water. There have also been spectacular water savings by individual firms through the adoption of recycling (Bhatia *et al.*, 1993).

Education and persuasion

Exhortation and appeals to public-spiritedness are often used as a temporary device, capitalising on public concern over droughts. In other cases they become a permanent feature of policy. Tariff increases are frequently part of a comprehensive package of measures, including public education and persuasion, and promotion of water efficiency (as in Bogor and Tucson). Disentangling the respective effects of the measures is difficult. In the most comprehensive assessment of the Tucson case (Martin *et al.*, 1988), the impact of non-price measures is played down, though this may understate their importance in preparing the public for price increases. There is little doubt that education and persuasion can strongly complement price measures, but they have very much less impact on their own.

To sum up, regulations and restrictions, if properly enforced, are predictable in their effects and their need can be readily understood by consumers. If fairly administered, they can be equitable in their impact between different socio-economic groups, and can penalise large and profligate users disproportionately. However, they depend on rigorous monitoring and enforcement. If implemented by weak and corrupt administrations, they tend to bear down with greatest severity on the less privileged groups. Nor do they necessarily maximise the benefits from water use, and may be excessive from the point of view of 'optimum' use (one of the conclusions of the Perth study in Hanke, 1982).

Exhortation, on the other hand, is much less certain in

its effects, though it has fewer political and administrative costs. It can, in the short term, help to bolster the impact of more rigorous measures (e.g. 'softening up' the public in advance of price increases) but its more permanent impact is doubtful.

PROJECTS AND PROGRAMMES

Most of the measures discussed so far in this chapter entail government action to create the framework for managing water demand, leaving individuals or firms to respond in their own best interests. The measures do not, on the whole, have major spending implications, nor do they entail sizeable direct government intervention in particular projects.

However, the impact of policies discussed earlier can often be reinforced by introducing specific public projects and programmes, so long as financial and administrative resources are adequate. At its most practical level, demand management includes direct intervention to improve the efficiency of the water delivery network, or programmes specifically to encourage user efficiency, recycling and re-use, etc.

The common element in these programmes is the aim of satisfying a certain demand for water services with reduced volumes of fresh water. From the viewpoint of supply managers, these measures reduce the demand for fresh water. However, from the consumers' standpoint, the same level of water services is provided from reduced bulk supplies. Whether this is regarded as supply-side or demand management is to a large extent a matter of viewpoint.

Interventions can occur at different points in the water system, and the distinction between supply and demand measures is not always easy to maintain. As one survey points out:

Distinctions between supply and demand are not

always consistent throughout the literature. The precise meaning of these terms depends on the point in the water delivery system where 'supply' is defined. [In this report] . . . supply will be defined at the entry point to the distribution system; after source, bulk storage, transmission and treatment works, but before distribution piping, distribution storage and customer taps

(Boland, 1991).

Another distinction is according to whether the prime mover is from the consumer or the supply side. The distinction is blurred in the case of deals between suppliers and consumers in their mutual interests. There are examples both of supply improvements undertaken by consumers in their own interests (e.g. urban agencies lining irrigation canals) and improvements in users' appliances encouraged by water utilities with a mandate to conserve (e.g. Tucson Water Company subsidising the installation of water-efficient toilets).

In this section we review evidence on three types of intervention, namely canal lining, leak detection, and the promotion of improved user efficiency.

Canal lining

A large proportion of water for irrigation is lost to farmers as a result of leaking from canals during transportation. Whether it is worth lining the canals to reduce these losses depends on a straightforward comparison of the costs of the programme with the benefits of the water conserved. Canal lining can be a very attractive solution to water supply problems.

In Bihar, India, lining the full length of a main irrigation canal was found to be economically justified. The distribution of benefits was also equitable, since it is the less influential farmers at the end of the system who tend to suffer from irregular supplies. A complication can

sometimes arise where leakage along a conveyor is necessary to replenish an aquifer, which may be of concern to
small farmers. In the Bihar case, however, the leakage
caused water-logging, so its reduction counted as a benefit
(Sinha and Bhatia, 1982).

In Southern California the Metropolitan Water District
of Los Angeles has come to an agreement with the
Imperial Irrigation District whereby the MWD bears
the cost of lining a main irrigation canal and purchases the
water saved for urban use. This was the least cost method
of obtaining the water in comparison with the main alternatives, and also had local environmental benefits (Wahl
and Davis, 1986).

Programmes to reduce unaccounted-for water

In many countries, so-called 'unaccounted-for water' is a
high proportion of supply, often reaching 25–50 per cent
of gross production in developing country networks
(Munasinghe, 1992). As we saw in Chapter 1, this category
is made up of technical losses such as leaks, under-metered
supplies, authorised non-paying customers (e.g. public
services) and deliberate theft (unofficial connections or
evasion of payment). The reduction of non-technical
losses is equivalent to tariff increases for these categories,
and the effects on benefits and welfare can be assessed
accordingly (Hanke, 1982). The case for reducing technical
losses is a straightforward matter of comparing the costs
of the programme with the value of water saved; it is
often a more attractive investment for a utility than the
creation of new supplies.

A programme in Sao Paulo, Brazil, reduced unaccounted-for water from 36 per cent to 31 per cent between
1980 and 1985. It consisted of the installation of meters,
leak detection, updating cadasters to discover which
houses had legal connections and which not, improving
maintenance and renovating old installations. Although a
modest proportion of total supply, the savings that

resulted were equivalent to the entire supply for a city of 2–3 million people (*The Urban Edge*, 1991).

Although such schemes are usually financially profitable, many water utilities prefer to invest in new facilities, since leak-detection programmes and improved billing procedures are complicated to implement. Hence, cost-benefit analysis of such programmes should make full allowance for the costs of administration and implementation. Otherwise, they have no direct impact on equity, and tend to be environmentally beneficial compared to the alternatives.

Improved user efficiency

Schemes seeking improvements in users' efficiency aim to squeeze more 'water services' out of a given volume of fresh water. One element is to promote water-efficient user appliances through a programme of demonstration and information, cheap loans and subsidies, and public education campaigns. In East Bay, California, large numbers of kits containing water-saving devices were distributed free. In both the Tucson and East Bay cases it is difficult to judge the effect of exhortation and water-saving campaigns since they were accompanied by tariff increases.

Consumers will be more attracted to seek savings where water tariffs are high enough to register as a significant cost. There is evidence from the energy sector that both households and firms have a very high required rate of return from efficiency measures. This may be partly irrational, but may also reflect the nature of the capital market in which they operate. Subsidies and tax breaks can help by shortening the pay-back period, and public exhortation and demonstration can spread information about the size of savings.

4

COMPARING THE OPTIONS

you will do your work on water
An' you'll lick the bloomin' boots of 'im that's got it.
(R. Kipling, *Gunga Din*)

The previous chapter referred to many instances where various kinds of demand-side measures were effective in conserving water and encouraging its reallocation to more socially valuable uses. This chapter discusses and develops the criteria that planners and project analysts can apply to this new breed of projects. The criteria are then applied to three typical, but differing, cases.

METHODOLOGY AND PERFORMANCE CRITERIA

Viewed as an alternative to supply augmentation projects, demand-management options typically require the use of different appraisal techniques. Supply-side options can be appraised using well-tried techniques of cost-benefit analysis (CBA) and/or cost-effectiveness analysis (CEA) (Munasinghe, 1992). Many schemes are justified by demonstrating least-cost methods of meeting project demand levels deemed to be fixed and unavoidable. Appraisals based on estimates of benefits can, in theory, use a full range of estimators (Gibbons, 1986), though in practice they tend to rely on estimates of willingness-to-pay

(Whittington, various), sometimes supplemented by quantified public health benefits.

Demand-management policies (apart from system improvements that leave the level of service unaffected) tend to involve a reduction in consumption and thus in the level of consumer welfare. The proper assessment of these policies thus entails predicting consumers' responses and placing economic weights on these changes. In theory, the effects of a demand reduction are symmetrical to those of an increase, and can be appraised by a simple analytical technique based on supply-demand analysis. This is reviewed below, along with other economic criteria such as the cost of conserved water.

The economic criterion is not the only one relevant to choosing the best ways of planning and managing water systems. A more complete list of criteria would include the following:

- efficacy
- economic efficiency
- equity
- environmental impact
- fiscal effects
- political and public acceptability
- sustainability
- administrative feasibility

In particular circumstances other criteria may also be relevant, e.g. impact on food self-sufficiency, regional development, the urban-rural balance, etc. These criteria are briefly discussed below.

Efficacy

All the policies examined in this book accomplished some conservation or reallocation – or, in the case of *ex ante* studies, were confidently expected to do so. However, policies varied in the amount of change they brought about, and in the extent to which they achieved their own declared objectives. A criterion important to policy-makers is the response to a given amount of policy 'effort', or the 'elasticity of response' to different policies.

The clearest measure of response is the elasticity of demand in respect of changes in the price of water. There is growing evidence that certain categories of demand are elastic, or less inelastic than others, such that price changes can induce demand responses. Even where demand is price-inelastic (less than 1.0), prices can still be effective in reducing consumption compared with other options for balancing supply and demand (see next section).

Virtually all the empirical work on elasticity, reviewed in Tables 4.1 and 4.2, is derived from developed OECD countries. Certain common elements can be observed. In-house domestic water consumption appears to be inelastic, whereas outdoor use is much more elastic. The greater demand elasticity of outdoor use also explains some of the seasonal and regional differences evident in the tables (Gibbons, 1986).

Low observed elasticities may reflect low water charges; where the cost of water is an insignificant part of the household budget, changes in price may not have a very dramatic effect. If water budgets were ever to become a significant part of household spending, a different and more elastic range of the demand function would probably come into play. In urban communities in developing countries with inadequate public service, consumers do pay high prices to private vendors, representing a sizeable part of their disposable income. In a study of Onitsha, Nigeria, it was found that one-third of households reported spending 10 per cent or more of their income on water in the dry season. The poorer segment of households, making up 58 per cent of the sample, spent an average of 18 per cent of their income on water (Whittington et al., 1991).

Water is a sensitive topic in most societies. Reforming public behaviour towards it is an invidious and difficult task with substantial political and administrative costs. It is therefore important that policies should have a

Table 4.1 Price elasticities for urban public water supply

Country	Location	Type of study	Estimated price elasticity	
Australia	971 households in 20 groups in Perth	readings over 1976–82; pooled x-section and time series	overall:	-0.11
Australia	315 households in Perth	x-section (hypothetical valuation technique)	in-house: ex-house: overall:	-0.04 -0.31 -0.18
Australia	metered	x-section	winter:	-0.36
Australia	137 households in Toowoomba Queensland	1972–3 to 1976–7 pooled cross-section and time series	short-term: long-term:	-0.26 -0.75
Canada	Urban demand eastern Canada	x-section 1960s	winter: summer:	-0.75 -1.07
Canada	Municipal demand Victoria, B.C.	time series 1954–70	winter: summer: mid-peak: year-round:	-0.58 zero -0.25 -0.40
England and Wales	411 firms in Severn-Trent	water-saving investment in 1972–78		-0.3
England and Wales	Industrial (metered) consumption England & Wales	time series 1962–80	year-round:	-0.3

Finland	Municipal demand Helsinki	time series 1970–78	year-round:	−0.11
Netherlands	Industrial demand, Rotterdam	time series 1960s and 1970s	'no price elasticity demonstrated'	
Sweden	69 domestic residences in Malmo	14 readings each over 1971–78; pooled cross-section and time series	year-round:	−0.15
United States	2,159 households in Tucson, Arizona (water use per household)	42 readings each over 42 months, July 1976–Dec. 1979; pooled cross-section and time series	year-round:	−0.256
United States	Domestic use in Tucson, Arizona	time series Jan. 1974–Sept. 1977	year-round (1): (log model) (linear)	−0.27 −0.45/−0.61
United States	Residential use in 21 study areas, eastern and western United States	cross section early 1960s	winter: summer: summer:	−0.06 (2) −0.57 (2) (east) −0.43 (2) (west)

1 Price included volumetric price of sewer use and the whole tariff schedule (increasing block was assumed to change in the same proportion as 'marginal rate' changes).

2 Changes in marginal price (= marginal block rate) only, although intramarginal rate structure allowed for in demand function. These elasticities represent significant reductions on those estimated from the same data fifteen years earlier (when the intramarginal rate structure was not allowed for): −0.23, −0.86 and −0.52, respectively.

Source: Herrington, 1987

Table 4.2 Consensus estimates of price elasticity of water demand in USA

	Short run elasticity	Long run elasticity
Residential use		
Indoor use	0.0	0.0 to −0.10
Outdoor use – Eastern US	n/a	−1.30 to −1.60
– Western US	n/a	−0.70 to −0.90
Commercial and institutional use		
Individual categories	n/a	−0.20 to −1.40
Industrial use (for water supplied from public system)		
Individual categories	n/a	−0.30 to −6.71
Aggregate industrial	n/a	−0.50 to −0.80

Source: Boland, 1991: 23

commensurate 'pay-off' in the effective fulfilment of their goals. Efficacy is thus related to the criterion of acceptability, discussed below. Typically, a combination of measures (e.g. regulation, tariff changes, subsidies for re-equipment) will be introduced, and it may be difficult to ascribe responsibility for the observed changes to any single one of them.

The discussion so far has been about the strength of demand response to price changes. For reallocation to be effected, there must also be the means for the physical transfer of water to take place. The water released must obviously be of a quality, in an amount, and in a location where it can be transported to other consumers with higher-value uses for it. In urban situations transferability is not normally an issue where the distribution system is well-developed. But for water saved in rural areas transfers may be hindered by geology and topography. For instance, the large amount of water used in agricultural areas to the south and east of Beijing would, if it were released, have to be pumped uphill to the city at some cost (Hufschmidt *et al.*, 1987).

Economic efficiency

The efficiency criterion requires that the economic benefits of policies exceed their costs, where both are discounted at an appropriate social discount rate. A simple
economic model that is useful in approaching water conservation is set out in Figure 4.1.

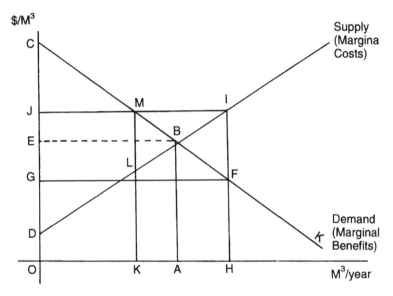

Figure 4.1 Supply and demand for water
Source: Herrington, 1987

The demand curve relates the consumer's willingness to
pay to the amount of water consumed. This would normally be downward-sloping from left to right, reflecting
the diminishing marginal valuation of successive
increments of water. The supply curve slopes upwards,
reflecting the fact that increments of demand can normally
be met only at rising cost to the water system. From the
point of view of fixing prices, the cost schedule is best
interpreted in the sense of long-run marginal costs of
expanding the system to meet a permanent increment
of demand. These basic notions are equally applicable to

water 'mining', such as the excessive drawdown of aqui-fers, where the benefit of reduced consumption is the avoided cost of alternative future supplies.

In Figure 4.1, net benefits are maximised when OA units of water are produced with a price of OE. Net benefits, the excess of the area under the demand curve over that under the supply curve, are represented by the area CBD. If consumption is higher than this, say OH, the costs ABIH of supplying the increment AH exceed benefits ABFH by BIF. Conversely, if consumption is restricted to OK, for instance by excessive prices or over-zealous restrictions, the loss of consumer benefits KMBA exceeds the supply cost savings KLBA, and this solution is also sub-optimal.

Various refinements to this basic model can be made:

1 Where there are marked seasonal patterns of demand and supply, a separate set of demand and supply curves would apply to each season, with the implication that there should, in theory, be different seasonal pricing.

2 Where a water supply system has unused capacity there is an important difference between short-term and long-run marginal costs. In the short run, with spare capacity, the extra cost of supply is likely to consist of the extra expenses of treatment and pumping, which are often minor. In the long term, new capacity must be created. In this situation, the ideal pricing in the short term – low enough to induce full take-up of capacity – would conflict with that based on long-term considerations.

3 Conservation includes the notion of increased efficiency in use, measured as the amount of water saved in pro-viding a given level of services. Such measures as reduced leakage from customers' taps and pipes, more efficient toilet flushes, etc. need not entail any loss of consumers' surplus. It would be misleading to describe it in terms of a movement along, or shift in, a standard water demand schedule since the relationship between price and demand for water is the same as before. Stric-

tly, the axis should be redefined as the demand for a given level of service rather than a physical amount of water.

4 Conservation arising from the reduction of unaccounted-for water commonly brackets together both physical losses of water and the forfeit of sales receipts by the utility. Reducing the former is a clear welfare gain, whereas the latter is a financial and distributional issue for the status of the utility (though also affecting its long term efficiency), and its relation with its customers.

5 The model has the greatest validity for a modern urban system in which the water utility has a monopoly and most users have piped connections. In other systems, water may be re-used several times, and intakes may be co-located with wastewater outlets. In these complex systems, the benefits of conservation will depend on the size of consumptive and non-consumptive use, and what happens to saved water. Conservation by users with a high consumptive element would release proportionately more water for the system than the same amount of conservation by users with a high non-consumptive use. The impact of conservation by, for example, firms using large volumes of water for cooling would result in less water being released, as well as less being taken in. The net effect on water available for downstream users or in-stream purposes would depend on the configuration of the system and the quality of water released. A further complication with conservation programmes for high-consumptive users such as agriculture is that their water losses may have benefits for other users (e.g. leakage from irrigation channels adding to groundwater supplies).

6 Where water supply is from a finite source, such as an aquifer being tapped beyond its recharge rate, the true cost of the water is the future cost of substituting for it when the aquifer is exhausted or contaminated. These costs, arising in the future, are discounted to obtain their present value. Units of water sold now thus bring

forward the date at which the aquifer is depleted. This depletion factor should enter into their price.

A version of this method has been applied to Manila, where depletion of the aquifer is projected to lead to its exhaustion over the next 26 years. Allowing this to happen would progressively raise the costs of pumping, and eventually costs of supply would stabilise at a higher level, equal to the cost of supplying water from new sources outside the city. In order to estimate the true costs of this eventuality, a conservation scenario was drawn up, in which the extraction of groundwater was deliberately reduced until the level of sustainable withdrawals was reached, and then maintained.

This alternative option has higher short-term costs, but long-term cost savings, compared to the depletion alternative. The difference in the discounted costs of the two scenarios is akin to a depletion cost, estimated to be US cents 1.2 per cubic metre in this case – as it happens, a rather small amount (Munasinghe, 1990).

Cost-benefit analysis can be supplanted or supplemented by cost-effectiveness analysis (CEA). CEA is applicable where benefits cannot be adequately measured, but where they are presumed to exist. CEA is also useful in comparing alternative, or cumulative, ways of attaining a given level of benefits. CEA can yield the discounted economic costs of achieving a unit of conservation.

Figure 4.2 contains the results of an exercise into cost-effective ways of meeting Beijing's future water requirements without major investments in new supply sources. It was discovered that one-third of industrial water consumption could be saved by the adoption of three measures: more recycling of industrial cooling water, recycling of power plant cooling water, and wastewater recycling, ranked in this order on a discounted basis. They were all substantially cheaper than the obvious next project to develop supply. In the domestic sector, it was found that four techniques could save 15 per cent of consumption,

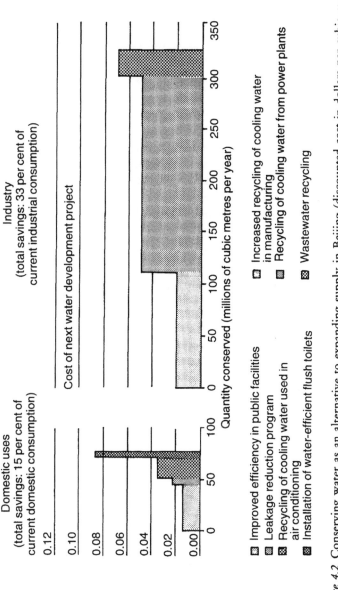

Figure 4.2 Conserving water as an alternative to expanding supply in Beijing (discounted cost in dollars per cubic metre)
Source: Hufschmidt *et al.*, 1987

and each of them was cheaper than the alternative of augmenting supply. These were: improving conservation in public facilities, programmes for the reduction of leakage, recycling air conditioning cooling water, and installing water-efficient flush toilets. If the (discounted) costs and amount of water saved by these measures are arrayed as in the table, they form a 'supply curve' of conserved water.

Equity

The conceptual model used so far assumes that the conservation of water results in less being supplied, in which case the benefits of this policy arise from savings in supply costs. Where water is *reallocated* a different notion of benefits is required – one which allows for the benefits derived by the new users of water.

Reallocation to higher-value uses produces net social benefits corresponding to the difference between the values of water in its old and new uses. There are, in principle, various ways of measuring this (Gibbons, 1986). Reallocation may also confer health benefits, e.g. when it leads to better provision for deserving social groups, or where it is associated with improved irrigation practices that reduce the habitat of waterborne disease vectors or reduce contaminated run-off. Equally, major reallocations may have local and regional external effects which need not be symmetrical with those of localities gaining from the transfer.

Reallocation can be envisaged as a shift in the structure of demand relative to recent historical patterns, brought about in response to policy changes. Reallocation would occur either within a sector, such as agriculture, or between sectors, such as from agriculture to urban use. Where reallocation happens in the context of trading between willing buyers and willing sellers in a free and undistorted market, our presumption is that the change would be from lower- to higher-value applications, since the price offered by the buyer would have to be sufficient to compensate for the seller's loss of benefit, plus transaction costs (ignoring for simplicity any environmental cost of the transfer). Where transfers occur in a distorted market, the financial benefits of the use will depart from the social value. A common instance would be distorted relative crop prices and subsidised inputs in irrigated farming.

Where reallocation is through administrative devices, such as quotas or edict, the water will not necessarily go into uses with a higher economic value, though the reallocation may serve social purposes. The users who have less water will be forced to adjust their consumption in this way, but those who gain more water can make the opposite adjustment – towards lower-value uses. The net effect could be either better or worse, and would depend partly on the quality of information available to the agency making the decision, as well as the motives for the shift.

Distributional effects

One criterion for policies is their relative impact on the various socio-economic groups. There are several aspects to this question. One is that deserving groups, such as women, poor households and small farmers, previously receiving supplies which were inadequate or obtained at high personal or social cost, should benefit from the reallocation, or as a minimum should not find themselves worse off. The consumption of such target groups should not, in any case, be reduced to below socially desirable levels.

Poorer groups in society, with less influence and voice, tend to get low priority in the public provision of water services. Poorer farmers are often at the tail-end of irrigation systems, where supplies are unreliable. Poorer urban consumers tend to be last in the queue for getting piped supplies and sewerage. Where conventional policies for water supplies often fail the poor, demand-management measures may be helpful in comparison. For instance, the poor may pay less for piped and metered supplies, at an economic tariff, compared with what they now pay to private vendors.

A related concern is that more affluent consumers should not receive disproportionate benefits from any policy measures, and that extreme inequalities in water consumption should be reduced.

Environmental impact

Environmental effects of the various options should, in principle, be included in the appraisal. In practice, only certain effects can be quantified, and even those only partially and imperfectly. Nevertheless, environmental effects should be factored into the economic appraisal, either as costs or credits, using recognised techniques (Winpenny, 1991, etc.). This has rarely been done until recently. This omission favours supply-augmentation options, which tend to have significant environmental costs, whereas the demand-management policies considered here avoid such costs and often have pollution-reducing benefits.

The approach taken in this book is to regard quantifiable environmental costs as elements in economic costs, to be taken into account in cost-benefit analysis and price-fixing. To the extent that the environmental effects have been undervalued or non-quantifiable, 'the environment' should be a separate criterion, along with the others mentioned above.

Fiscal effects

Many countries with serious water problems also have weak public finances. The fiscal impact of water policies is important, both for general macroeconomic management and for the proper funding of water and sanitation provision. The primary criterion is the net effect of a policy on the finances of central or local government, whichever is the more relevant. The secondary criterion is the effect on the finances of the water utility, irrigation agency, etc. These effects can be either positive or negative, e.g. the gain to the budget from a tax, price increase or charge, or the costs to the budget from a subsidy or tax relief. The impact of the approach proposed here would normally be to strengthen the finances of central government and the water utility.

The literature has dealt extensively with the public finance implications of adopting marginal-cost pricing (e.g. Turvey, 1968). In the common situation of increasing long-term marginal costs, this pricing principle would generate 'excessive' revenue for the water utility compared to the alternative of average-cost pricing. This revenue could breach allowable rates of return established by regulatory bodies, and could be embarrassing in other ways. Such an effect could be neutralised by adjusting consumer charges which are unrelated to consumption, e.g. the fixed part of a two-part tariff, or by reducing the price of the first 'blocks' in an 'increasing block' tariff structure (Zamora et al., 1981).

The fiscal yield of a specific price adjustment depends on the price elasticity of demand. Although elasticities vary greatly for different categories of consumption, most aggregate estimates have values less than 1.0 (Boland, 1991: 23). Where this is the case, tariff increases will increase total revenue.

Political and public acceptability

Policies that are acceptable to the parties affected have better prospects of being implemented than those which are likely to encounter serious resistance in the political process, or amongst consumers or other interested parties. There will normally be some proportionality between the effort that goes into introducing a policy measure (sacrificing political goodwill, spending political credit, steering legislation through, overcoming public resistance and lobbying, etc.) and the pay-off from that policy. A policy that achieves little, but at great political cost and arousing much public antagonism, is clearly undesirable.

Some of the factors that determine acceptability are: the severity of the problem; the distribution of costs/benefits; the lead given by political and community figures; the publicity; the educational level and public-spiritedness of the population; and the extent of the behaviourial changes implied (as opposed to a technocratic fix).

Sustainability

Certain policies have a once-and-for-all impact, while others have a continuing or even a growing effect. Short-term measures introduced in response to an emergency, such as a drought, may have a strong immediate impact, but one which tails off sharply when the worst of the emergency is over. Policies which make a long-term impression on water use, such as technological adaptations and changes in user habits, are more sustainable. Best of all are measures whose impact increases over time, e.g. because their elements reinforce each other, or because they provide incentives for continuing and cumulative effects.

Administrative feasibility

This has several facets. Operating a policy must be within the administrative capability of the department or agency involved. Its effects should be monitorable with the resources of the responsible agency, and the regulatory agency should have adequate powers of enforcement. Policies should be intelligible to those affected.

APPLICATIONS

The use of the above criteria for comparing and choosing water policies can now be illustrated by some actual cases, exemplifying various different approaches to demand management. These include tariff increases combined with a vigorous conservation campaign (Bogor, Indonesia), the purchase by an urban authority of surplus irrigation water, combined with a canal lining programme (MWD/IID, California), and plans by a heavy industrial company to conserve and recycle water (TISCO, in Jamshedpur, India). These cases are described more fully in Bhatia *et al.*, 1993. Other cases are the subject of current research by the author.

Water tariffs and household demand in Bogor, Indonesia

Bogor is a large town in west Java with a population of approximately 250,000. Most of its water is drawn from springs close by, supplemented from a river and small deepwells. Almost half the population is served by the water supply company, PDAM, mostly from household connections. Per capita average consumption is 169 litres per day. Available supply is 420 litres/second (l/s), almost a quarter of which is unaccounted-for. In 1987 the average charge for a residential consumer using 30 cubic metres per month was only Rp 108/cubic metres, compared to the unit cost of production of around Rp 440.

On official projections, the population of Bogor is expected to grow to 1.35 million by 2010, and PDAM's plans are to extend coverage to 87 per cent of this number, providing 132 l/head. These supplies could best be secured by exploiting surface sources, mainly the Cisadena River, and the first phase of a project to extract 940 l/s from this source has begun, at a cost of US$46.6 m. However, the longer-term project to increase the offtake from the river would cost around twice this (Rp 800 per cubic metre) and still leave a deficit of supply. It was therefore decided to combine supply augmentation with demand-management measures.

A new tariff structure was posted in 1988, raising the

rates by between 100 per cent and 280 per cent, increasing its progressivity, and raising fixed service charges by between 50 per cent and 275 per cent. The steepest increases in cost were for larger consumers, those using more than 30 cubic metres per month. A domestic user with a monthly demand of 30 cubic metres would pay Rp 6,000 per month, compared with the old amount of Rp 2,500.

As noted in Chapter 3, PDAM supplemented its tariff measures in 1989 with a conservation campaign aimed at the largest consumers – those with a monthly demand of more than 100 cubic metres. Pamphlets and brochures describing ways of reducing water use, and instructions on reading meters, were sent to every customer. For larger consumers, written information on possible reasons for excessive consumption was also sent, and home visits by PDAM officers were offered as a follow-up. For customers joining the Consumption Level Evaluation Programme, officers checked the premises for leaks, estimates of repair costs and savings were given, and advice on water use habits was provided.

Following the criteria developed in the previous section the programme can be assessed as follows.

Efficacy

Although an assessment of its long term impact is not available, the programme had a significant impact in its first nine months. In response to the tariff measures (the conservation campaign was not introduced until six months later) domestic and commercial water use fell by about 30 per cent. Three months after the conservation campaign started, average monthly water use fell by 29 per cent, and that of consumers previously using over 100 cubic metres per month fell by 26 per cent.

Economic efficiency

Applying the Hanke (1982) method to the Bogor demand data, the tariff increase can be shown to be economically efficient in that savings in resource costs from reduced consumption exceed the loss of consumer welfare from lower usage. Extending the same analysis to the conservation campaign, it can also be shown that the resource costs to the utility and household from house visits, leak repairs, etc. exceed the value of the decrease in consumption. On simplified assumptions for a 'standard' consumer, the cost of leak repairs could be balanced by a saving in water bills in around three months.

Equity

The unit price of water for small consumers (up to 10 cubic metres per month) is practically one-tenth that of large consumers (though the addition of fixed charges brings the total monthly bill up to around one-third of the latters'). As regards the large number of people (currently about half) still without a proper service, the prospects for a connection are likely to be much greater if existing amounts can be spread around, than if sole reliance has to be placed on costly new supply projects.

Environmental impact

The better use of a given volume of supply obviates or delays the need for new diversionary schemes, or aquifer pumping. The reduction of leaks is also likely to improve neighbourhood amenity.

Fiscal impact

Increased tariff revenue benefited the finances of the water utility, and postponement of major new investment eased the pressure to incur new debt.

Acceptability

Customers who joined the conservation programme were said to be pleased with the service from PDAM. Savings resulting from the campaign helped to offset the higher water bills.

Sustainability

Provided tariffs are regularly reviewed and adjusted as necessary, the incentive for careful water-use habits can be maintained.

Administrative feasibility

The prior existence of comprehensive metering implies a sizeable and efficient workforce to read and maintain the meters and run the billing service. But once in existence, the cost of adjusting tariffs becomes very low. The conservation campaign, on the other hand, made intensive use of skilled and semi-skilled employees, though it appears in this case that PDAM successfully coped with the demand, and did itself much credit.

Water conservation agreement in Southern California

For most of this century Los Angeles has been engaged in plans to secure its long-term water supply. A succession of major investment projects (notably the Los Angeles, the Colorado River and the California Aqueducts) has left the city still projecting a supply deficit for a normal year by the year 2000. This would arise partly through the expected growth in the city's population, and partly because current supplies from the Colorado River and from Northern California through the State Water Project are likely to fall, because of increased consumption by users with prior rights. The deficit in 2000 was, in the mid-1980s, estimated to be around 140,000 acre-feet (Wahl and Davis, 1986).

The Metropolitan Water District (MWD) examined four main options, any of which would have bridged the projected deficit:

1 Building reservoirs in Central California as part of the State Water Project, and pumping water south along existing aqueducts. This was reckoned to be the most expensive solution, with unit costs of $310–440 per acre-foot, depending on the options chosen.
2 Enlarging the Shasta Dam on the federal Central Valley Project, and pumping water south along existing aqueducts. Although relatively cheap ($285 per acre-foot), such a deal would have to recognise the prior rights of farmers using federal water, which would introduce some uncertainty into supply.
3 Raising the price of water to MWD consumers in order to curb demand and enable existing supplies to stretch further. However, public resistance was expected, and the increased revenue for the MWD could have violated its statutory revenue ceiling.
4 Introducing water conservation measures in the Imperial Irrigation District (IID) and transporting the savings along the Colorado River Aqueduct to the MWD.

The last option was chosen. Around a third of all water delivered to the scheme from the Colorado was lost to productive use. The large irrigated area (450,000 acres) contained an extensive and complex network of canals, drains, gates, and spillways from which extensive losses were incurred, and the main conveyor, the All-American Canal, was earth-lined. Farmers themselves had little incentive to conserve water when it only cost $9 per acre-foot. The MWD agreed to invest $125 million in water conservation in the IID service area, which was expected to yield savings of 100,000 acre-feet per year, to be transferred to MWD for 35 years. This is 4 per cent of all water used annually by the IID.

The investment was to go on a mixture of conservation works and programmes, such as canal lining, interceptor canals, providing reservoir-regulating capacity, centrally controlled automated gates at major canal junctions, installing tail-water recovery systems, and allowing farmers to call for water in shorter (12 hour) periods. (Wahl and Davis, 1986; Keller *et al.*, 1992).

The agreement can be assessed as follows.

Effectiveness

The IID should have little difficulty in delivering the agreed amounts of water to the MWD. Since the water originates in the Colorado, the IID has not suffered the severe curtailment of supplies that has affected other parts of California. Although there are doubts about the yield of the canal-lining programme, there are a number of other conservation opportunities so far not covered by the agreement which could be drawn upon (e.g. on-farm conservation measures – Keller *et al.*, 1992).

Economic efficiency

On the criterion of the cost of the conserved water, the MWD/IID agreement is one of the lowest, and does not suffer the uncertainty over supplies inherent in the cheapest option – enlarging the Shasta Dam. The water would be transferred from low-value uses in irrigated agriculture (worth $30–35 per acre-foot) to higher-value consumers in the MWD. The transfer also neatly deals with growing drainage costs incurred by the IID. No farmers are likely to use less water than before, hence there is no sacrifice of consumer surplus.

Equity

There are no obvious equity concerns. There are no losers from the deal, and the entire water-consuming population of Los Angeles obtains water at a cost lower than that of alternatives.

Environmental impact

The agreement avoids the construction of new dams, reservoirs or conveyances that would be necessary for some of the alternative solutions. No new structures are required. There is even an environmental bonus in that the level of the inland Salton Sea had been rising due to drainage flows from the IID, and this effect would be lessened. The one possible problem is that the quality, as opposed to the quantity, of water flowing into the Salton Sea could become worse if tail water – which is cleaner than drainage – is diverted to the MWD.

Fiscal impact

The arrangement with the IID will save the MWD outlays on more expensive alternatives. The option of raising tariffs would have greater benefits but, as already noted, would have been embarrassing *ipso facto*.

Acceptability

The benefit of the transfer was clear to the major parties involved. No existing farmers were likely to get less water. Reducing leaks and wastage is totally uncontroversial, provided it is achieved at an acceptable cost (which in this case was set at $125 per acre-foot). The deal even saved the IID some embarrassment since its farmers have been the only ones in California to have escaped the dire effects of the recent drought, which has devastated their competitors. The IID was also under strong pressure to reduce the environmental damage caused by its high level of wastewater flows into the Salton Sea, which was costing it money in lawsuits.

Sustainability

Provided the structures and programmes financed under the deal are maintained, the savings should become a permanent feature. In any case, other possible water savings could be tapped.

Administration

Once the civil works are undertaken, the task of operating the system to realise the savings falls on existing IID staff. Maintaining the lining on canals to achieve the target rate of loss reduction is a difficult task, as is operating tailwater recovery systems (Keller *et al.*, 1992).

Industrial conservation in Jamshedpur, India

Thanks to the operations of the Tata industrial group, Jamshedpur is a rare example of a city 90 per cent of whose water for household, commercial and industrial purposes is provided by the private sector. The city's industrial sector (13 major units, 16 medium enterprises, and 567 small concerns) accounts for 55.5 million cubic metres per annum, while 65 million cubic metres per annum is treated and distributed to domestic and institutional consumers in a population of about 1 million. The water is almost all drawn free of charge from the River Subernarekha through a system operated by the private Tata Iron and Steel Company.

This free source is under threat. As part of the Subernarekha Multipurpose Project financed by the World Bank, a reservoir is being built near the city to serve an irrigation area of 82,000 hectares with a potential demand of 720 million cubic metres per annum. It is almost inevitable that the Irrigation Department will start charging the city for its water, and the spectre of competition for limited supplies will arise. As a result, industry in Jamshedpur has started examining ways of conserving water and treating effluent and municipal sewage for recycling in industrial processes.

The Tata Iron and Steel Company (TISCO) itself is the largest user, drawing 33 million cubic metres per annum, and it is planning to double its steel production capacity. The firm estimates that it could reduce its water consumption per unit of steel by 40 per cent over two years, saving 13 million cubic metres per annum. The capital cost of achieving these savings is not currently available. However, costings are available for the conservation programme of the Tin Plate Company, the third largest industrial water user, which is planning to enlarge its productive capacity. Its options are as follows:

1 To invest in water conservation processes; reduce water consumption by 5 million cubic metres per annum (compared with 6.64 million cubic metres if nothing is done); and reduce the investment cost of effluent treatment, since the volume of wastewater will be much less.
2 Make no investment in water conservation; pay a higher water bill on the full 6.64 million cubic metres per annum; and invest in larger effluent treatment plant.

There are trade-offs between investment in conservation and investment in effluent treatment, and between investment in conservation and the bill for water. The costs of the two options are as follows:

Option 1: Rs140m for the conservation process; Rs30m for effluent treatment; annual operating costs Rs1m.
Option 2: Rs100m for the effluent treatment plant; annual costs Rs22m.

On an annualised basis, the extra cost of option 1 is Rs9.6m (US$370,000). This is the (discounted unit) cost of saving 5 million cubic metres per annum of water, equivalent to US$70 per thousand cubic metres, or 7 US cents per cubic metre.

From the viewpoint of the private firm, whether or not it is worth investing in the conservation process depends on the (future) financial cost of water. At a tariff of 7c. per cubic metre or more, the investment is worthwhile.

From a social point of view, getting firms to invest in water-conserving processes is one amongst a number of options to increase available supplies. Other options include the use of treated and recycled effluent and sewage, and the development of new supply systems. These options can be ranked in ascending order of unit cost to produce a water supply curve (e.g. that for Beijing in Figure 4.2), and ideally implemented in the same order. The costs should, if possible, include some element of the environmental costs (or give credit for benefits) associated with each option (e.g. using treated municipal sewage saves the cost of public sewage treatment). In principle, these options should be undertaken up to the point where the cost of the marginal unit supplied/conserved equals the value of that unit. This can be obtained from a demand curve, as in Figure 4.1.

The option of encouraging private firms to invest in processes that conserve water can be assessed as follows.

Effectiveness

Once the investment in process changes is made, the water savings are secure; – they do not depend on consumer behaviour. Over time, it is likely that the competition for water in this locality would encourage firms to redouble their efforts to economise on water, and methods which may not be profitable now (e.g. using treated effluent and sewage) may well become so in future.

Economic efficiency

Releasing water through private industrial conservation is in this case likely to be cheaper than public investment in new supply sources. Whether the savings are worthwhile in an absolute sense depends on the marginal value of the water saved, on which data are lacking in this case.

Equity

Industrial conservation has strong equity implications. A proportion of the population in the Jamshedpur metropolitan area is currently unserviced, and an additional number receive a meagre service from the municipalities. Improving services to these groups would be greatly aided if water could be available without costly public investments. If, in future, companies also started using treated effluent and sewage, contamination of the river – which is currently very high – would be reduced, and it could become available as a source of fresh water for the unserviced urban population.

Environmental impact

By avoiding or postponing investment in new supply sources, conservation avoids potential environmental costs.

Fiscal impact

If conservation is carried out without public subsidy, it avoids any public spending implications. In the event that conservation was not worthwhile to the firm at a given financial tariff, there could be a case for public subsidy. Where there were public benefits (externalities) to be obtained from the production of water – for instance if water could be released to agricultural users with higher marginal values – there would be a case for public subsidy.

In this case it would be cheaper for the authorities to provide water for agriculture by subsidising industry to release it by conservation.

Acceptability

Investment by private firms in their own interests is, to the general public, an invisible method of securing extra water supplies. Since heavy industry is a major water consumer, it is to be expected that it should take a lead in economising on scarce supplies.

Sustainability

Permanent changes in production technology are an assurance of water savings. In order to keep the pressure on industrial and commercial users to continue the search for water efficiency, water prices (or, in certain cases, effluent charges) need to be continually reviewed to provide the required incentive.

CONCLUSION

This chapter has proposed and developed eight criteria to apply to water supply solutions: efficacy, economic efficiency, equity, environmental impact, fiscal effect, acceptability, sustainability and administrative feasibility. These criteria have been applied to three varied examples of demand-management, in Indonesia, California and India. The cases all stand up well on these criteria, and support the argument for systematically including demand-management options alongside supply-augmentation solutions to water-stressed cities and regions.

5

CONCLUSIONS
The urgency of reform

With earth's waters make accord
(F. Thompson, *To a poet breaking silence*)

This book has indicated an urgent need for a drastic
change in the way we view and use water – recognising
its value to others, and its potential scarcity, and treating
it more on a par with other goods and services. This
implies radical shifts in attitudes, policies, institutions and
technology. The evidence of Chapter 3 is that, faced with
problems approaching critical severity, communities devise
solutions or expedients to fit. Many of these solutions
involve market mechanisms: the invisible hand is, so to
speak, reaching up from the grass-roots even if it is not
in evidence reaching down from the macroeconomy. The
crucible of reform is not in planning ministries, but in
city and regional water departments. Changes are very
much demand-driven, rather than the product of planning
scenarios.

Could water problems therefore take care of themselves,
without needing the intervention of planning models or
major new public investments? After all, water is, unlike
biodiversity or the ozone layer, a renewable resource and
today's mistakes do not – in general – lead to irreversible
consequences. Why not leave the problems for local communities to sort out?

This attitude is too glib. It overlooks circumstances

where the abuse of water does have results which are, in effect, irreversible – such as the depletion of an underground aquifer to the point where it is contaminated with saline water or polluted with heavy metal residues. It also makes light of the large social and economic costs entailed in the transition to more rational solutions, such as the large populations without adequate safe supplies, and communities whose normal water source is depleted or polluted.

In short, the ground needs to be carefully prepared for reforms to succeed. The policy tier described in Chapter 3 should, ideally, be put in place. Nothing will succeed, however, without a change in attitudes by all involved. The United Nations Conference on Environment and Development in Brazil in 1992 signalled the importance of treating water as a scarce economic resource, in the following terms:

> A prerequisite for the sustainable management of water as a scarce vulnerable resource is the obligation to acknowledge in all planning and development its full costs. Planning considerations should reflect . . . investment, environmental protection and operating costs, as well as the opportunity costs reflecting the most valuable alternative use of water . . . The role of water as a social, economic and life-sustaining good should be reflected in demand management mechanisms.
>
> (UNCED, 1992, Chapter 18, Agenda 21, paras 8.16/17)

This final chapter recaps the main arguments of the book and highlights some of the implications for policy reforms. The final section touches on the delicate art of political economy, and considers how the changes could be brought about.

THE ARGUMENT RESTATED

Lack of coherent policies

The water sector is often fragmented, and rarely planned and managed as a whole. Responsibilities for providing supply and disposal services to different groups of consumers tend to be divided amongst several agencies, with their own investment plans and operating and pricing criteria. The value of water varies widely between different users, a *prima facie* sign of sub-optimal resource use.

Policy failures

The traditional approach to water provision has relied mainly on 'supply augmentation'. Investment programmes have been aimed at satisfying a projection of 'reasonable' requirements. The role of price has, at the most, been seen as recovering some of the cost of current provision. Water tariffs tend to be based on average costs, and often are a great deal lower. Water prices do not, in general, reflect the resource cost of provision, as represented by the Long Run Marginal Cost (LRMC). Consequently, from an economic point of view, there is 'excessive' consumption, much of it for low-value purposes.

The same is true of water pollution. The failure to charge the economic rate for fresh water, and to penalise pollution, leads to contamination of surface and groundwater bodies. Both the provision of water and its use and disposal impose environmental costs on others which are not reflected in prices charged to the users.

The social and environmental benefits from satisfying minimum standards of water use in the population can be acknowledged in special 'lifeline' rates for certain social groups and for minimum blocks of consumption, and by cross-subsidising sewerage and sanitation from water provision.

The minimum conditions for an optimal economic allocation of water would be to price it at its true resource cost by applying LRMC pricing, and to 'internalise' its environmental costs in prices and charges borne by the consumer. The growth of water markets would also allow the alternative use values of water to find some reflection in its price, which would help to improve its allocation.

This approach was endorsed by the Dublin Conference (1992) in the following terms:

> Past failure to recognise the economic value of water has led to wasteful and environmentally damaging uses of the resource.

In practice, this philosophy implies an emphasis on demand management, as opposed to supply augmentation, and the greater use of water prices and markets to provide the right signals to suppliers and users. An important element is to improve the use of existing supply by encouraging conservation and reallocation, with the aim of equating marginal values with resource costs. The urgency of such an approach is evident in the rapidly increasing costs of investment in new sources of supply and the growing signs of water shortage and pollution in many developing countries.

Efficacy of markets and prices

Markets and prices are effective in producing sizeable savings of water in households and industry, provided they are used boldly and in the context of an enabling environment. This evidence is opposed to much conventional opinion, which argues that price elasticities are too low to make these policies effective. Water markets and banks have been instrumental in transferring water between farmers, and between agriculture and industrial and urban users. Conservation measures, sometimes involving a package of non-market and market instruments, have produced savings of 20–30 per cent or more.

Importance of enabling conditions

The greater use of the market is a necessary but not sufficient condition for the more rational allocation of water. Market forces will be frustrated unless other elements are present in the policy framework. Enabling conditions include legal and institutional reforms affecting the water sector, and the creation of a macroeconomic context in which prices can function effectively.

Market incentives may need to be reinforced by non-price measures, such as legal restrictions (especially in controlling pollution), quotas and norms. Private monopoly suppliers may need to be brought under public regulation. Public support for the policy package needs to be enlisted through adequate publicity and education campaigns, though there is evidence that these are ineffective without pricing measures or fiscal penalties. The impact of price measures can be further assured by direct programmes of conservation, such as canal-lining, leak detection, the subsidised distribution of efficient appliances, etc.

Environmental bonuses

Compared to supply augmentation projects, demand management can be attractive in economic *and* environmental terms. (There is a close analogy to the power sector, where energy efficiency measures compare favourably with investments in new power schemes.) Conserving and reallocating existing water supplies avoids the high economic and environmental costs of developing new supplies, and of coping with the disposal of more wastewater. Using the 'cost of conserved water' criterion, conservation is frequently cheaper than investing in alternative supply sources. Greater recycling and re-use on the part of industries and utilities likewise reduces the need for fresh water, and also tend to reduce effluent pollution.

Equity

Conservation and reallocation measures are often beneficial in their effects on poorer consumers. In a situation typical of many cities in developing countries, only the more affluent parts of the population are served with adequate piped water supplies and there is a large and growing backlog of investment to extend the system to poor areas. Meanwhile the poor depend either on standpipes of dubious reliability or buy water at a high unit cost from private distributors. In any case they bear the brunt of supply restrictions and failures. It is instructive to compare this representative picture of public supplies with that in cities relying on private services, where coverage can be more universal and service standards higher.

Solutions relying on heavy investment in new sources will have little impact on the overall supply shortage, given the rapid growth of urban populations and the pervasive funding problems of water utilities. But increases in the efficiency of water use, the reduction of waste and leakage, and conservation by large users such as industry would release relatively large amounts for redeployment elsewhere. The analogous situation in agriculture is the plight of poor farmers at the end of the irrigation system, who tend to be the first to suffer shortages when demands exceed the system's ability to deliver. Canal lining, improvements in the efficiency of water use and attention to leakage and waste should improve the quality of service to those at the end of the chain.

The more active pricing of water services need not penalise the poor. Many poor urban households already pay high unit rates for water – much higher than those paid by residents with piped supplies – and accounting for a sizeable part of their budgets. (Whittington et al., 1991). Reallocating water from low-value uses, such as in agriculture, is an obvious way of providing for the growth of poor urban populations and the industrial and commercial sectors on which they depend for employment. The alternatives would entail higher costs which would have to be passed on in tariffs.

Tariff structures can – and do – incorporate low rates for minimum levels of consumption, as a means of shielding poorer, or smaller, users from hardship, and encouraging minimum levels of consumption on social grounds.

IMPLICATIONS OF POLICY REFORM

These conclusions imply certain changes in the way water problems are approached. On a positive note, the following seem desirable:

1 Better use should be made of existing water supplies before major new supply schemes are planned and implemented. This implies an active use of policies for conserving and reallocating water.

2 Water sector policy needs to be shaped within a policy framework constructed in three tiers: enabling conditions, incentives, and direct interventions.

3 The policy framework should encourage local and regional initiatives to solve water problems, by making institutions flexible and responsive. The use of market devices should be part of this endeavour.

4 Within the relevant planning and policy-making area (whether catchment, conurbation, district, etc.) solutions to fresh water supplies, the disposal of wastewater, and water pollution should be sought in an integrated manner, where appropriate.

5 The 'water sector' needs to be viewed as a whole for planning, policy-making and investment purposes. The traditional compartmentalised approach, reflected, for instance, in separate agencies for irrigation, municipal supply, and pollution control, encourages the sub-optimal use of scarce water, and a failure to recognise externalities in water use and disposal. It will, however, become increasingly important for policy-makers, regulators, utility managers and others to take a 'holistic'

109

view of the sector to avoid the waste and conflicts that arise from uncoordinated behaviour. This is not, however, the same as advocating comprehensive, integrated national investment planning and modelling.

6 Measures of economic (including environmental) costs and benefits should be developed for appraisal and pricing purposes.

7 Among the opportunities for public spending which are consistent with the approach advocated in this book are: supply-side projects for redesigning systems and reducing wastage; leak detection programmes; extending the coverage of meters; canal lining, in some situations; provision of water-efficient and recycling equipment to public facilities and state-owned installations; and, the promotional financing of user efficiency devices.

The systematic implementation of these demand-side measures will facilitate the urgent task of extending adequate and safe water supply, sanitation and wastewater disposal to the large and growing number of people without proper provision. This is rightly highlighted as one of the most important environmental and development problems to be tackled in the 1990s (World Bank, 1992).

THE ART OF REFORM-MONGERING

It is safe to predict that turning water into a commodity will prove controversial and will draw deeply on a government's goodwill and credibility with its citizens. Governments should, however, seek every opportunity to remind their citizens of the alternative, which, in many cases, will be a grim scenario of growing and eventually disastrous water stress.

Drawing on experience of successful, as well as unsuccessful, reform programmes, we can point to the following *desiderata* to facilitate public acceptance of the necessary changes.

1 Exploit complementarities and create synergy between

110

the different elements in a reform programme; and create virtuous circles. The profitability of unaccounted-for water (UFW) programmes is increased if tariffs are set at realistic levels. Consumers will more readily espouse water-efficient appliances if tariffs are at economic levels. They will be more ready to pay higher tariffs if they see evidence of improved services. Industrialists are more ready to pay pollution fines and charges if the funds are earmarked for visible environmental clean-ups. A vigorous public campaign stressing the value of water and the dire consequences of allowing present trends to continue should form part of any reform programme.

2 Create gainers as well as losers. The most effective way of doing this is to promote water markets. Farmers who decide to sell their water instead of using it on low-value crops are doing themselves, as well as society, a favour. There are many interests to be mobilised in favour of pollution penalties and reduced industrial water use: one firm's effluent is someone else's intake.

3 Make the reforms socially equitable. In developing countries the poor are among the worst victims of existing systems. They regularly pay prices per unit for their water (from private vendors) many times higher than those paid by wealthier people with their own connections. Any reform that raises charges, improves cost recovery and generates funds for expanding and improving the system promises to be socially equitable, even if charges to piped consumers are raised. The structure of tariffs can further promote distributional goals by offering low 'lifeline' rates for minimum levels of consumption.

4 Exploit environmental benefits. The better management of demand will postpone, or even obviate, the need for investment in new supply – with its attendant environmental costs. Reduced water stress will also alleviate user conflicts – between municipalities and farmers, power utilities and fishermen, industrial polluters and

111

recreationists, as well as between nations. A reformed water sector confers environmental, as well as economic and financial, bonuses.

5 Enlist private resources. Government has dominated the planning, supply, distribution and disposal of water. This has happened for a mixture of motives, both noble and ignoble. Outright privatisation is not feasible or even desirable in every case, though it is a way of deflecting heavy future costs – and public obloquy – onto the private sector (as in the UK). Privatisation stopping short of a transfer of ownership (e.g. concessions, management contracts, contracting out) can combine the advantages of public ownership with private management. But the most thoroughgoing 'privatisation' is the devolution of reforms to private water consumers themselves. The more active use of tariffs and effluent charges will encourage firms and households to re-examine their water-using habits and invest in conservation. Promoting water markets and raising tariffs decentralise the task of matching demand with supply, and mobilise every party behind solving the problem. Nothing less will suffice.

BIBLIOGRAPHY

Agarwala, R., 1983: 'Price distortions and growth in developing countries'. World Bank Staff Working Paper No. 575, Washington, D.C.

Allan, Tony, 1992: 'Fortunately there are substitutes for water: otherwise our hydropolitical futures would be impossible', paper in ODA, 1992.

Anderson, Terry L. and Leal, Donald R., 1988: 'Going with the flow: expanding the water markets'. Cato Institute Policy Analysis No. 104. April 26.

Arlosoroff, Saul, 1985: *Water Management in Arid Zones*. Ivory Coast.

Batstone, Roger and Kosmo, Mark, 1989: 'Industrial pollution in the Mediterranean'. World Bank/EIB Environmental Program for the Mediterranean, Working Paper 5.

Bernstein, Janis D., 1991: 'Alternative approaches to pollution control and waste management: regulatory and economic instruments'. Discussion Paper of UNDP/World Bank/UNCHS Urban Management Program. Washington, D.C.

Bhatia, Ramesh, Cestti, Rita and Winpenny, James, 1993: *Policies for Water Conservation and Reallocation: Good Practice Cases in Improving Efficiency and Equity*. World Bank, Washington, D.C.

Bhatia, Ramesh, and Falkenmark, Malin, 1991: 'Water resource policies and the urban poor: innovative thinking and policy imperatives'. Paper prepared for Dublin International Conference on Water and the Environment, January 1992.

Boland, John, 1991: 'Legislative and economic approaches to

water demand management'. United Nations, New York, 1991.

Caswell, Margret, Lichtenberg, Erik, and Zilberman, David, 1990: 'The effects of pricing policies on water conservation and drainage'. *American Journal of Agricultural Economics 72*, November.

Cestti, Rita, 1989: 'Water resources: problems and issues for the water supply and sanitation sector'. Unpublished paper, INUWS, World Bank, Washington, D.C., August.

Charney, Alberta H. and Woodard, Gary C., 1990: 'Socio-economic impacts of water farming on rural areas of origin in Arizona'. *American Journal of Agricultural Economics 72(5)*, December.

Clarke, Robin, 1991: *Water: The International Crisis*. Earthscan Publications Ltd, London.

Coe, Jack J., 1988: 'Responses to some of the adverse external effects of groundwater withdrawals in California', in O'Mara, ed., 1988.

Colby, Bonnie G., 1990: 'Transaction costs and efficiency in western water allocation'. *American Journal of Agricultural Economics 72(5)*, December.

Coyaud, Daniel P., 1988: 'Private and public alternatives for providing water supply and sewerage services'. Discussion Note INU 31, World Bank Infrastructure and Urban Development Department, Washington, D.C.

Crosson, Pierre R., Cummings, Ronald G., and Frederick, Kenneth D., 1978: *Selected Water Management Issues in Latin American Agriculture*. Resources for the Future/Johns Hopkins University Press, Baltimore and London.

Darling, Arthur H., Gomez, Christian, and Niklitschek, Mario E., 1992: 'The question of a public sewerage system in the Caribbean: a case study'. Paper presented by the Inter-American Development Bank at the CIDIE Workshop on 'Environmental economics and natural resource management in developing countries', World Bank, January.

Department of Water Resources (DWR), 1991: *California's Continuing Drought 1987–1991*. State of California, December.

Dinar, Ariel and Letey, J., 1991: 'Agricultural water marketing, allocative efficiency and drainage reduction'. *Journal of Environmental Economics and Management*, 20, 210–233.

Dublin Conference, 1992: International Conference on Water

and the Environment, Dublin, January. Quotations from the 'Dublin statement on water and sustainable development'.

El-Ashry, Mohamed T., and Gibbons, Diana C., 1986: *Troubled Waters: New Policies for Managing Water in the American West*. World Resources Institute, Washington, D.C., October.

Eskeland, Gunnar S., and Jimenez, Emmanuel, 1991: 'Choosing policy instruments for pollution control: a review'. Working Paper, WPS 624. World Bank, Washington, D.C., March.

Falkenmark, Malin, 1989: 'The massive water scarcity now threatening Africa – why isn't it being addressed?', *Ambio* 18(2).

Falkenmark, Malin, Garn, Harvey and Cestti, Rita, 1990: 'Water resources: a call for new ways of thinking'. *Ingenieria Sanitaria* XLIV, Enero-Junio.

Findley, Roger W., 1988: 'Pollution control in Brazil'. *Ecology Law Quarterly*, 15(1).

Fisher, Anthony C., 1981: *Resource and Environmental Economics*. Cambridge University Press.

Franceys, Richard, 1990: 'Paying for water – urban water tariffs'. *Waterlines* 9(1) (Nigeria).

Frederick, Kenneth D., 1991: 'The disappearing Aral Sea'. *Resources* 102. Resources for the Future, Washington, D.C.

Frederick, Kenneth D., ed., 1986: *Scarce Water and Institutional Change*. Resources for the Future, Washington, D.C.

Gelb, Alan H. and Gray, Cheryl W., 1991: *The Transformation of Economies in Central and Eastern Europe: Issues, Progress and Prospects*. World Bank, Washington, D.C.

Gibbons, Diana C., 1986: *The Economic Value of Water*. Resources for the Future, Washington, D.C.

Golladay, Fredrick L. and Katsu, Shigeo [undated]: 'The role of prices in management of water resources'. Unpublished paper. World Bank, Washington, D.C.

Grubb, Michael, 1990: *Energy Policies and the Greenhouse Effect*. Royal Institute of International Affairs, London/Dartmouth Publishing Co, USA.

Gupta, Devendra B. and Bhatia, Ramesh, 1991: 'Water conservation through pricing and pollution control: a case study of two fertilizer plants in India'. Unpublished paper.

Hanke, Steve H., 1982: 'Economic aspects of urban water supply: some reflections on water conservation policies'. Collaborative

Paper, CP–82–91 of the International Institute for Applied Systems Analysis, Austria.

Hanley, N., Hallett, S., and Moffatt, I., 1990: 'Why is more notice not taken of economists' prescriptions for the control of pollution?'. *Environment and Planning A* 22.

Hardin, G., 1968: 'The tragedy of the commons'. *Science*, 162, 1243–8.

Harrigan, Jane, Mosley, Paul & Toye, John, 1991: *Aid and Power: The World Bank and Policy-based Lending*. Routledge, London.

Herrington, P., 1987: *Pricing of Water Services*. OECD, Paris.

Howe, Charles W., Schurmeier, Dennis R., and Shaw, William D., 1986: 'Innovations in water management: lessons from the Colorado-Big Thompson Project and Northern Colorado Water Conservation District', in Frederick, ed., 1986.

Hufschmidt, Maynard, Fallon, Louise, Dixon, John, and Zhu, Z., 1987: *Water Management Policy Options for the Beijing-Tianjin Region of China*. East-West Center, Hawaii.

Kanbur, Ravi, 1990: 'Projects versus policy reform', in Supplement to *The World Bank Economic Review* and *The World Bank Research Observer*.

Katko, Tapio, 1988: 'Pricing of water services in Finland and some other developed countries'. *Aqua Fennica* 18(1).

Katko, Tapio S., 1990: 'Cost recovery in water supply in developing countries'. *Water Resources Development* 6(2), June.

Keller, Jack, Peabody, N.S.III, Seckler, David, and Wichelns, Dennis, 1992: *Water Policy Innovations in California: Water Resource Management in a Closing Water System*. Center for Economic Policy Studies, Winrock International Institute for Agricultural Development, Arlington, Va., January.

Kennedy, David N., 1991: 'Allocating California's water supplies during the current drought'. A discussion paper presented at the World Bank's International Workshop on Comprehensive Water Resources Management Policies, Washington, D.C., June.

Killick, A., 1991: 'Notes on macroeconomic adjustment in the environment' in *Development Research: The Environmental Challenge*, ed. James T. Winpenny, ODI, London.

Kinnersley, David, 1991: 'The UK's experience of water privatization'. Paper contributed to a seminar of the World Bank's Water and Sanitation Division, May.

Kneese, Alan V., 1984: *Measuring the Benefits of Clean Air and Water*. Resources for the Future, Washington, D.C.

Kosmo, Mark, 1989: 'Economic incentives and industrial pollution in developing countries'. Environment Department Division Working Paper No 1989–2, World Bank, Washington, D.C., July.

Kotlyakov, V. M., 1991: 'The Aral Sea Basin: a critical environmental zone'. *Environment* 33(1).

Leitch, Jay A., and Ekstrom, Brenda L., 1989: *Wetland Economics and Assessment: An Annotated Bibliography*. Garland Publishing, New York and London.

Leonard, H. Jeffrey, 1988: *Pollution and the Struggle for World Product: Multinational Corporations, Environment and Comparative Advantage*. Cambridge University Press.

Lovei, Laszlo, and Whittington, Dale, 1991: 'Rent seeking in water supply'. Discussion Paper, Report INU 85, Infrastructure and Urban Development Department, World Bank, Washington, D.C.

Macrae Jr., Duncan and Whittington, Dale, 1988: 'Assessing preferences in cost-benefit analysis: reflections on rural water supply evaluation in Haiti'. *Journal of Policy Analysis and Management* 7(2).

Magrath, William, 1990: 'The challenge of the commons: the allocation of non-exclusive resources'. Environment Department Working Paper, World Bank, Washington, D.C.

Manu, Yvonne, 1991: 'Back to the office report on a visit to Madras'. Internal document, Water and Sanitation Division, World Bank, Washington, D.C.

Martin, William E. and Kulakowski, Susan, 1991: 'Water price as a policy variable in managing urban water use: Tucson, Arizona'. *Water Resources Research* 27(2), February.

Martin, William E., Ingram, Helen M., Cory, Dennis C., and Wallace, Mary G., 1988: 'Toward sustaining a desert metropolis: water and land use in Tucson, Arizona', in *Water and the Arid Lands of the United States*, ed. M. T. E. Ashry and D. C. Gibbons, Cambridge University Press, New York.

Michelsen, Ari M., and Young, Robert A., 1989: 'Economics of optioning water rights for urban water supplies during drought'. Paper presented to 25th Annual Conference of the American Water Resources Association, September.

Miglino, Luis Caetano, 1984: 'Industrial wastewater management in Metropolitan Sao Paulo'. Ph.D. thesis, Harvard University.

Moncur, James E.T., 1987: 'Urban water pricing and drought management'. *Water Resources Research* 23(3), March.

Moore, Michael R., 1991: 'The Bureau of Reclamation's new mandate for irrigation water conservation: purposes and policy alternatives'. *Water Resources Research* 27(2), February.

Munasinghe, Mohan, 1990: 'Managing water resources to avoid environmental degradation: policy analysis and application'. Environment Working Paper No. 41, World Bank, Washington, D.C., December.

Munasinghe, Mohan, 1992: *Water Supply and Environmental Management*. Available from: Munasinghe Volume, 4201 East West Highway, Chevy Chase, MD 20815, USA.

National Rivers Authority (NRA), 1991: *Scheme of Charges in Respect of Applications and Consents for Discharges to Controlled Waters*. Solihull, UK.

Nickum, James E., and Easter, K. William, 1990: 'Institutional arrangements for managing water conflicts in lake basins'. *Natural Resources Forum*, August 1990.

Nisbet, Ian C.T. and Risebrough, Robert W., 1989: 'Pollution problems in the Mediterranean: approaches and priorities'. World Bank/EIB Environmental Program for the Mediterranean, Working Paper 1.

OECD, 1987: *Pricing of Water Services*. Paris.

OECD, 1989: *Water Resource Management: Integrated Policies*. Paris.

OFWAT, 1990: *Paying for Water*. Office of Water Services, Birmingham, UK.

Okun, Daniel A., 1991: 'A water and sanitation strategy for the developing world', *Environment* 33(8), October.

O'Mara, Gerald T., ed., 1988: *Efficiency in Irrigation: The Conjunctive Use of Surface and Groundwater Resources*. World Bank, Washington, D.C.

O'Mara, Gerald T., 1988, 'The efficient use of surface water and groundwater in irrigation: an overview of the issues', in O'Mara, ed., 1988.

ODA, 1992: 'Priorities for water resources allocation and management'. Proceedings of a conference at Southampton, July. Overseas Development Administration, London.

Pezzey, John, 1990: 'Charge-subsidies versus marketable permits as efficient and acceptable methods of effluent control: a property rights analysis'. Discussion Paper No. 90/271. University of Bristol, Department of Economics.

Phantumvanit, Dhira, and Panayotou, Theodore, 1990: *Industrialization and Environmental Quality: Paying the Price*. Thailand Development Research Institute, Bangkok.

Portney, Paul R., ed., 1990: *Public Policies for Environmental Protection*. Resources for the Future, Washington, D.C.

Postel, Sandra, 1989: *Water for Agriculture: Facing the Limits*. Worldwatch Paper 93, Washington, D.C.

Randall, Alan, 1981: 'Property entitlements and pricing policies for a maturing water economy'. *The Australian Journal of Agricultural Economics* 25(3), December.

Randall, Alan, 1988: 'Market failure and the efficiency of irrigated agriculture', in O'Mara, ed., 1988.

Reidinger, Richard, 1992: 'Observations on water markets for irrigation systems'. Paper for World Bank's Ninth Annual Irrigation and Drainage Seminar. Washington, D.C., December.

Reisner, Marc, 1990: *Cadillac Desert*. Secker & Warburg, London.

Repetto, Robert, 1986: *Skimming the Water: Rent-seeking and the Performance of Public Irrigation Systems*. World Resources Institute, Washington, D.C.

Rhodes, George F., and Sampath, Rajan K., 1988: 'Efficiency, equity and cost recovery implications of water pricing and allocation schemes in developing countries'. *Canadian Journal of Agricultural Economics* 36, 103–117.

Robinson, Robert S., and MacDonnell, Lawrence J., 1990: *The Water Transfer Process as a Management Option for Meeting Changing Water Demands*. Report for the US Geological Survey, April.

Rogers, Peter, 1990: 'Concept paper for World Bank Comprehensive Water Resources Management Policy Paper'. Unpublished, Harvard University. July 19.

Roth, Gabriel, 1987: *The Private Provision of Public Services in Developing Countries*. World Bank/Oxford University Press.

Roumasset, James, and Chakravorty, Ujjayant, 1988: 'An economic approach to water-use planning'. Chapter 1 of 'Efficiency principles for water management' by Roumasset, Chakravorty, Wilson and Moncur. Working Paper No. 2, Environment and Policy Institute, East-West Center, Hawaii.

119

Roy, B.K., 1991: 'Water availability in India'. *Water Resources Development* 7(2), June.

Saliba, Bonnie Colby, 1987: 'Do water markets "work"? Market transfers and trade-offs in the Southwestern States.' *Water Resources Research* 23(7), July.

Saliba, Bonnie Colby, and Bush, David B., 1987: *Water Markets in Theory and Practice: Market Transfers, Water Values and Public Policy.* Westview Press, Boulder and London, 1987.

Saliba, Bonnie Colby, Bush, David B., and Martin, William E., [undated]: *Water Marketing in the Southwest – Can Market Prices be Used to Evaluate Water Supply Augmentation Projects?* General Technical Report RM-144. US Department of Agriculture Forest Services. Fort Collins, Co. 80526.

Seagraves, James A., and Ochoa, Renan, 1978: 'Water pricing alternatives for Canete, Peru', in Crosson, Cummings and Frederick, ed., 1978.

Sexton, Richard, 1990: 'Perspectives on the Middle East water crisis: analysing water scarcity problems in Jordan and Israel'. ODI/IIMI Irrigation Management Network Paper 90/3f.

Shah, Tushaar, 1985: 'Transforming ground water markets into powerful instruments of small farmer development: lessons from the Punjab, Uttar Pradesh and Gujerat'. Irrigation Management Network Paper 11d, ODI, London, May.

Shah, Tushaar, 1989: 'Efficiency and equity impacts of groundwater markets: a review of issues, evidence and policies'. Research Paper 8, Institute of Rural Management, Anand, India. November.

Simon, Benjamin, and Anderson, David, 1990: 'Water auctions as an allocation mechanism in Victoria, Australia'. *Water Resources Bulletin* 26(3), June.

Sinha, Basawan and Bhatia, Ramesh, 1982: *Economic Appraisal of Irrigation Projects in India.* Agricole Publishing Academy, New Delhi.

Small, Leslie E., and Carruthers, Ian, 1991: *Farmer-Financed Irrigation: The Economics of Reform.* Cambridge University Press.

Sunman, Hilary, 1992: 'The application of charging schemes for the management of water pollution: experience and prospects'. Environmental Resources Limited, London.

Triche, Thelma A., 1990: 'Private participation in the delivery of Guinea's water supply services'. Working Paper WPS 477.

Infrastructure and Urban Development Department, World Bank, Washington, D.C.

Triche, Thelma, 1991: 'Seminar on privatization of water supply in the UK and the role of the private sector in France, September 1991. Summary of the two approaches and discussion of issues'. Internal World Bank report, November.

Turvey, R., ed., 1968: *Public Enterprise*. Penguin Books, UK.

UNCED, 1992: *Agenda 21*. Document of the United Nations Conference on Environment and Development. Rio de Janeiro.

UNIDO, 1990: *Industry and Development*. Annual Report of the United Nations Industrial Development Organisation, Vienna.

United Nations, 1991: *Criteria for and Approaches to Water Quality Management in Developing Countries*. Natural Resources Water Series No. 26. United Nations Department of Technical Cooperation for Development, New York.

The Urban Edge, April 1991, World Bank, Washington, D.C.

Vaux, Henry J., Jr., 1986: 'Economic factors shaping Western water allocation', *American Journal of Agricultural Economics*, December.

Vaux, Henry J., Jr., 1991: 'The California drought, 1987–?'. *Newsletter* of Water Science and Technology Board, National Research Council 8(2). April.

Wahl, Richard W., and Davis, Robert K., 1986: 'Satisfying Southern California's thirst for water: efficient alternatives'. Chapter in Frederick, 1986.

Warford, Jeremy J., 1968: 'Water supply', in Turvey, R., ed.

Waterbury, John, 1988: 'Legal and institutional arrangements for managing water resources in the Nile Basin'. In O'Mara, ed., 1988.

White, Judy, 1991: 'Quantification of the benefits of rural potable water supply projects'. Draft ODA guidance note, unpublished, Overseas Development Administration, London.

Whittington, Dale, 1991: 'A note on the development of a USAID water resources policy for developing countries'. Unpublished, July.

Whittington, Dale, Mu, Xinming and Roche, Robert, 1990: 'Calculating the value of time spent collecting water: some estimates for Ukunda, Kenya'. *World Development* 18(2).

Whittington, Dale, Lauria, Donald T., Okun, Daniel A., and Mu,

Xinming, 1989: 'Water vending activities in developing countries: a case study of Ukunda, Kenya'. *Water Resources Development* 5(3), September.

Whittington, Dale, Lauria, Donald T., and Mu, Xinming, 1991: 'A study of water vending and willingness to pay for water in Onitsha, Nigeria'. *World Development* 19(2–3).

Whittlesey, Norman K., 1990: 'The impacts and efficiency of agriculture-to-urban water transfer: discussion'. *American Journal of Agricultural Economics* 72(5), December.

Wilczynski, Piotr, 1990: 'Environmental management in centrally-planned non-market economies of Eastern Europe.' Environment Department Working Paper No. 35, World Bank, Washington, D.C., July.

Winpenny, James T., 1991: *Values for the Environment: A Guide to Economic Appraisal.* Her Majesty's Stationery Office, London.

Winpenny, James T., 1992a: 'Powerless and thirsty? Prospects for energy and water in developing countries'. *Utilities Policy,* Special Edition, October, 1992.

Winpenny, James T., 1992b: 'Water as an economic resource'. Paper delivered to a Conference of the Overseas Development Administration, University of Southampton, July.

Wooldridge, R., ed., 1991: *Techniques for Environmentally Sound Water Resources Development.* Pentech Press, London.

World Bank, 1992a: *World Development Report 1992.* Washington, D.C.

World Bank, 1992b: *Water Supply and Sanitation Projects: The Bank's Experience 1967–89.* Washington, D.C.

World Bank/EIB, 1990: *The Environmental Programme for the Mediterranean.* World Bank, Washington, D.C. and the European Investment Bank, Luxembourg.

Young, Robert A., 1986: 'Why are there so few transactions among water users? *American Journal of Agricultural Economics,* December.

Zamora, Jennifer, Kneese, Alan V., and Erickson, Erick, 1981: 'Pricing urban water: theory and practice in three Southwestern cities'. *The Southwestern Review* 1(1), Spring 6.

INDEX

Note: All references are to water, unless otherwise specified.

79; policy mix 46–8, 69, 71, 73
Turkey 42
Turvey, R. 89

UFW *see* unaccounted-for water
unaccounted-for water (UFW: leaks, wastage and thefts) 5, 6, 16, 26, 111; and performance criteria 82, 83, 86–7, 97; and policy mix 40, 49, 72–3; urgency of reform 107, 108
underpricing *see under* tariffs
United Kingdom 112; comparing options 78; policy mix 33–4, 37, 44, 52, 67; problem 6, 7, 18
United Nations Conference on Environment and Development (1992) 104
United States: comparing options 79–80, 94, 95, 96; education and persuasion 69; enabling environment 34, 35, 37, 40; problem 5, 6, 11, 19; projects and programmes 71, 72, 73; restrictions and legal sanctions 67; surface water markets 57–61; transferable water-use permit 65–6; water banks 62–5; water tariffs 44–50; *see also* California
unpaid bills *see* unaccounted-for water
urban areas 108; performance criteria 77–9, 82–4, 91–102; and policy mix 34, 38–9, 58,

59–60; problem 4, 6, 17; *see also* households
urgency of reform 2, 103–12; enabling environment 107; environmental bonuses 107; equity 108–9; implications of 109–10; lack and failure of coherent policies 105–6; markets and prices, efficacy of 106; reform-mongering 110–12; *see also* policy mix
Uttar Pradesh 55
Uzbekistan 5

Vaux, Henry J. Jr. 62
Victoria (BC, Canada) 78
Victoria (State, Australia) 55, 56
volumetric measurement of water consumption 43–4, 50–1

Wahl, Richard W. 72, 95, 96
Warford, Jeremy J. 9
wastage *see* unaccounted-for water
wastewater (run-off, effluent and sewage) 9, 12, 26; *see also* pollution; sanitation
water *see* comparing options; policy mix; problem; tackling problem; urgency of reform
'water stress' 3, 6–7
water-intensity sectors *see* agriculture; industry
water-saving appliances 73
West Africa 24, 37–9
West Bank 3
wetlands, loss of 4, 13

Milton Keynes UK
Ingram Content Group UK Ltd.
UKHW040013071024
449327UK00011B/207